International Conference on Polymeric, Metallic and Composite Materials (PMCM 2021)

IOP Conference Series: Materials Science and Engineering
Volume 1167

Wuhan, China
29 - 30 May 2021

ISBN: 978-1-7138-3431-1
ISSN: 1757-8981

Printed from e-media with permission by:

Curran Associates, Inc.
57 Morehouse Lane
Red Hook, NY 12571

Some format issues inherent in the e-media version may also appear in this print version.

Institute of Physics
Dirac House, Temple Back
Bristol BS1 6BE UK

Phone: 44 1 17 929 7481
Fax: 44 1 17 920 0979

techtracking@iop.org

Additional copies of this publication are available from:

Curran Associates, Inc.
57 Morehouse Lane
Red Hook, NY 12571 USA
Phone: 845-758-0400
Fax: 845-758-2633
Email: curran@proceedings.com
Web: www.proceedings.com

TABLE OF CONTENTS

IOP Conf. Series: Materials Science and Engineering 1167 (2021) 011001 doi:10.1088/1757-899X/1167/1/011001

Preface

Thanks to Prof. Tien-Chien Jen, the Chair Professor at Anhui Agriculture University and the Director of UJ-Nanjing Tech Joint Research Center in China, for assisting Modern International Green Cultural Communication Association (MIGCCA) to successfully hold the 2021 International Conference on Polymeric, Metallic and Composite Materials (PMCM 2021) during May 29-30, 2021 in Wuhan, China.

In light of the worldwide outbreak of COVID-19, there were only 20 people present at the meeting. In such a small academic conference, all participants have more opportunities to communicate and build relationships with other scholars. At the meeting, Prof. Abd El-Fatah Abomohra, as the keynote speaker, made a speech to discuss the potential uses of microalgae in human health, wastewater treatment and biofuels. Dr. Kinjal J. Shah, who was not available to attend the conference in China, sent us a prerecorded talk, which made the form of participation more diversified. Meanwhile, we highly encouraged doctorate (PhD) and postgraduate students to present their research proposal and findings in this conference. With the concerted efforts of us all, the conference has been a complete success and contributed significantly to promoting dialogue and exchanges among scholars and researchers to report most recent innovations and developments, summarize state-of-the-art, and exchange ideas and advances in the aspects of polymeric, metallic and composite materials.

In addition, the PMCM 2021 has received 35 manuscripts, 17 submissions have been accepted by the reviewers. The acceptance rate is about 48.6%. The total number of reviewers involved is up to 45. All submitted full papers have been peer-reviewed by two reviewers. The acceptance would be granted if the recommendations from the reviewers were positive. The criteria are based on the technical criteria, quality criteria, presentation criteria and format. Authors should make sure the submitted papers use the template for this conference.

On behalf of the PMCM 2021 Organizing Committee members, we express our deep appreciation to all authors, speakers and participants of this conference for their contributions. Particular thanks also go to all the referees for their constructive comments on all papers who coordinated the linguistic review of the selected papers with competence and professionalism. This collection of selected papers offers an very interesting panorama of original theoretical and applied developments, representing key topics of polymeric, metallic and composite materials.

We'll look forward to seeing you next year at the next PMCM conference!

PMCM 2021 Organizing Committee
May 31, 2021

List of Committees

General Chair
Prof. Tien-Chien Jen
Anhui Agriculture University, China

Program Chair
Prof. Omid Rahmani
Faculty of Engineering, University of Zanjan, Iran

Committee Chair Members
Prof. Marcin Marek Kamiński
Lodz University of Technology, Poland

Prof. Mahdi Ghassemi Kakroudi
Department of Material Science and Engineering, University of Tabriz, Iran

International Technical Committee Members

Dr. Junhui Hu
Nanjing University of Aeronautics and Astronautics, China

Dr. Nabisab Mujawar Mubarak
Faculty of Engineering and Science, Curtin University, Malaysia

Dr. Bing Wang
School of Mechanical Engineering & Automation, Fuzhou University, China

Dr. Yasser Shahzad
Department of Pharmacy, COMSATS University Islamabad, Pakistan

Dr. Khalid Z. Elwake
Faculty of Science, Port-Said University, Egypt

Dr. Sasikumar Swamiappan
School of Advanced Sciences, Vellore Institute of Technology, India

Dr. Grzegorz Lesiuk
Wroclaw University of Science and Technology, Poland

Dr. Abhishek Gupta
University of Wolverhampton, Wolverhampton, England

Dr. Ehsan Kianfar
Chemical Engineering at Department of Chemical Engineering, Islamic Azad University, Iran

Dr. Hwai Chyuan Ong
University of Technology Sydney, Sydney, Australia

Dr. Erfan Babaee Tirkolaee
Istinye University, Istanbul, Turkey

Assoc. Prof. Karol Kyzioł
AGH University of Science and Technology, Poland

Peer review declaration

All conference organisers/editors are required to declare details about their peer review. Therefore, please provide the following information:

- **Type of peer review: Single-blind / Double-blind / Triple-blind / Open / Other (please describe)**

Double-blind peer review: All submitted full papers have been peer-reviewed by two reviewers. The acceptance would be granted if the recommendations from the reviewers were positive. The criteria are based on the technical criteria, quality criteria, presentation criteria and format. Authors should make sure the submitted papers use the template for this conference.

- **Conference submission management system:**

Submit to the conference mailbox as an attachment

- **Number of submissions received:**

35 submissions received

- **Number of submissions sent for review:**

35 submissions were sent for review

- **Number of submissions accepted:**

17 submissions are accepted

- **Acceptance Rate (Number of Submissions Accepted / Number of Submissions Received X 100):**

48.6%

- **Average number of reviews per paper:**

Average two reviews per paper

- **Total number of reviewers involved:**

45 reviewers

- **Any additional info on review process:**

The conference organisers evaluated each submission to determine if its topic and content are suitable for consideration before being reviewed. Manuscripts that do not meet the conference topic were returned to the authors.

Submissions that passed the initial review would be sent to two or three referees based on their expertise. The reviewers were asked to evaluate the manuscript based on its originality, significance, technical quality, relevance and completeness, and to complete their reviewing work (review form and revised draft) within two weeks. After collecting the referees' reports, the conference organisers would make a recommendation on the acceptability of the manuscript.

The final recommendation can be 1) Paper accepted for publication without changes (only in case at least 7 out of 13 criteria were evaluated positive and no criterion was evaluated negative); 2) Paper accepted for publication after modification; 3) Paper not eligible for publication (obligatory in case at least 7 out of 13 criteria were evaluated negative). If the paper requires corrections, the

IOP Conf. Series: Materials Science and Engineering 1167 (2021) 011002 doi:10.1088/1757-899X/1167/1/011002

reviewer would specify suggested remarks and comments for the author(s). Then the conference organisers would sent the remarks and comments to the author(s).

- **Contact person for queries (please include: name, affiliation, institutional email address)**

Emma FONG
Modern International Green Culture Communication Association (MIGCCA)
No. 888, Xiong Chu Road, Hongshan District, Wuhan, Hubei, China
icmed2017@outlook.com; emma@migcca.com

IOP Conf. Series: Materials Science and Engineering 1167 (2021) 012001 doi:10.1088/1757-899X/1167/1/012001

Effect of different steel fiber types on mechanical properties of ultra-high performance concrete

Xinyi Yan*, **Yuxin Gao, Yaoling Luo, Yao Bi and Yuhao Xie**

China West Construction Academy of Building Materials Co., Ltd., Chengdu 610015, China

*Corresponding author's e-mail: yanxy@cug.edu.cn

Abstract. The influence of steel fiber on the mechanical properties of ultra-high performance concrete (UHPC) were studied through slump, compressive strength, flexural strength test. The experimental results showed that the incorporation of steel fiber had an influence on the working performance of UHPC, and the slump and expansion degree of UHPC were inversely proportional to the volume content and length-diameter ratio of steel fiber. The steel fiber significantly improved the compressive and flexural strength of UHPC. The mechanical property of UHPC was the best when the steel fiber content was 4%, with 28 day compressive strength up to 163.2MPa and flexural strength up to 45.9MPa. UHPC showed a trend of gradual decrease of compressive strength and gradual increase of flexural strength with the increase of aspect ratio of steel fiber on the condition of constant volume content of steel fiber. The addition of special-shaped steel fiber can improve the compressive and flexural strength of UHPC effectively.

1. Introduction

Ultra-high performance concrete (UHPC) was a new type of cement-based material characterized by ultra-high strength, ultra-high toughness and ultra-high durability. Since its appearance, ultra-high performance concrete had attracted wide attention of researchers [1-3]. However, with the increase of strength, the brittleness of concrete was more obvious [4]. Therefore, steel fiber was gradually added into concrete to improve the performance of brittleness and burst failure, and enhanced the mechanical properties of matrix concrete, such as compressive strength, tensile strength and flexural strength [5-6].

In recent years, many scholars had studied the influence of steel fiber on the mechanical properties of concrete. Yan [7] found that adding appropriate amount of steel fiber can significantly improve the brittleness and toughness of ultra-high strength concrete. Yoo [8] and Yang [9] studied the effect of steel fiber volume fraction on the mechanical properties of concrete. The results showed that the difference of steel fiber volume fraction would hinder the reinforcement effect of steel fiber on matrix concrete. Song [10] had carried on the confirmation in the experiment to the textile fiber coordination supposition through to experimented the mixture ratio,the result showed that simultaneously the textile fiber when two kinds of fibers mixed most greatly to mix the quantity to have to be higher than textile fiber most greatly to mix the quantity, explained textile fiber mixed when had the textile fiber synergistic effect. By randomly dispersing steel fiber into UHPC, Kim [11] and Rong [12] found that the "skeleton frame" effect of fiber could effectively inhibit the development of micro cracks and the formation of macro cracks in concrete,then the mechanical properties of UHPC could be improved effectively. Meng [13] and Wu [14] studied the effect of steel fiber content on the shrinkage performance of UHPC, and the results showed that the incorporation of appropriate amount of steel fiber could effectively alleviate the local stress concentration and reduce the shrinkage. However, excessive fiber would lead to an increase in the porous weak interface area, which would reduce the shrinkage inhibition. Banthia [15] found that the bent hook steel fiber with complex shape can provide better end anchorage effect in the process of concrete cracking.

The use of steel fiber to enhance the strength and performance of concrete was an important means of concrete modification. After recognizing the excellent properties of steel fiber, this paper explored the regularity of steel fiber reinforced UHPC by means of macroscopic test method. The mechanical properties of UHPC with different steel fiber types under the influence of dosage and morphology were compared and analyzed, and the strengthening mechanism was discussed.

2. Materials and methods

2.1. Preparation of the basic materials

Table 1. Physical and mechanical properties of the steel fiber

Steel Fiber	Diameter/mm	Length /mm	Aspect Ratio	Tensile Strength /MPa	Shape
SF8	0.2	8	40	2900	Straight
SF13	0.2	13	65	2900	Straight
SF16	0.2	16	80	2900	Straight
SF40	5	40	8	2900	Wave
SF25	0.2	25	125	2900	Straight end hooked
SF30	0.2	30	150	2900	Wave end hooked

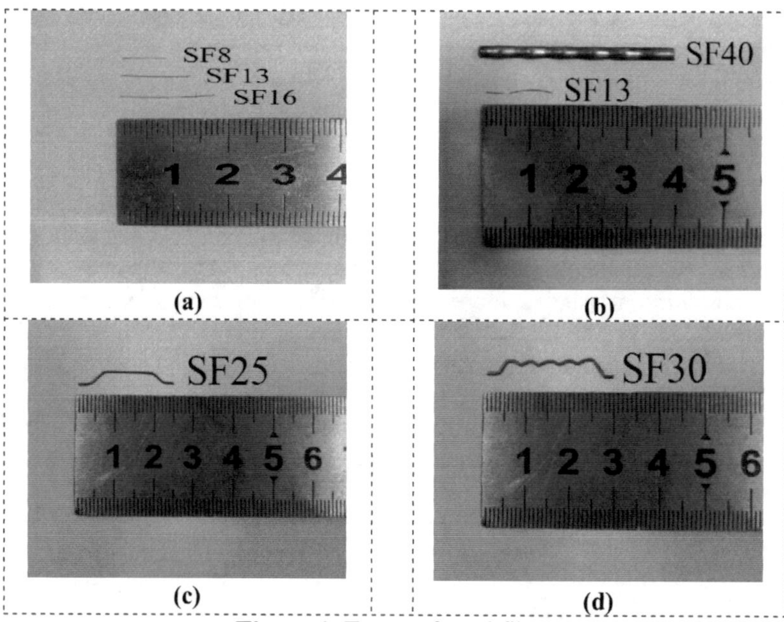

Figure 1. Types of steel fiber

The ordinary portland cement 52.5R, semi-encrypted grade 94 silica fume and S95 granulated blast furnace slag were applied in this paper.

Fine aggregate: natural river sand was used after screening and grading, and the fineness modulus of natural river sand was 1.3.

Chemical admixture: the admixture for UHPC with high water reduction, viscosity reduction and shrinkage reduction functions which is adopted.

Steel fiber: copper steel fiber, the main material property parameters of the six steel fibers are shown in the Table 1. The six different shape types are shown in the Figure 1.

2.2. Mix proportion

By changing the volume ratio or type of steel fiber, the effect of volume ratio of steel fiber and structure on the physical and mechanical properties of UHPC was studied. According to the mix proportion of UHPC shown in the Table 2 to mould the 100mm × 100mm × 100mm and 100mm × 100mm × 400mm samples.

Table 2. Mix proportion (kg/m^3) of the prepared UHPC

Cement	Granulated Blast Furnace Slag	Silica fume	Fine aggregate	Cement-sand Ratio	Water-binder Ratio	Chemical admixture
790	105	158	1053	1:1	0.18	0.6%

2.3. Design of experiments

Test implementation of slump and expansion performance of concrete mixture was carried out in accordance with *Standard Test Method for Performance of Ordinary Concrete Mixtures (GB / T50080-2002)*. *Standard Test Method for Fiber Concrete (CECS 13-2009)* was implemented in the mechanical properties test of concrete.

3. Results and discussion

3.1. Effect of steel fiber on the flow performance of UHPC

The influence of steel fiber content on the flow performance of UHPC is shown in the Table 3. The large amount of steel fiber incorporation would reduce the fluidity of the mixture and the uniformity of the steel fiber distribution, thereby reducing the overall fluidity of UHPC. Steel fiber had the characteristics of fine diameter, high strength and hardness, which played a role of "skeleton" when mixed with concrete. The steel fiber could absorb the moisture in the matrix and prevent the concrete from flowing to a certain extent, which increased the compactness of the whole concrete after filling the voids.

Table 3. Effect of steel fiber content on the performance of UHPC

Number	Steel Fiber	Volume content /%	Slump /mm	Expansibility /mm
1	-	0	270	650
2	SF13	1	280	635
2	SF13	2	260	515
3	SF13	3	280	635
4	SF13	4	290	740
5	SF13	5	250	565
6	SF13	6	230	400
7	SF8	2	290	528
8	SF16	2	120	230
9	SF13:SF40	2:1	260	530
10	SF13:SF40	1.5:1.5	275	660
11	SF13:SF40	1:2	280	645
12	SF13:SF25:SF30	0:2:0	270	590

With the increase of steel fiber aspect ratio, the slump and expansion of UHPC decrease gradually. In the process of concrete pouring, steel fibers with large aspect ratio were more likely to be distributed in the direction perpendicular to the flow of concrete, hindering the flow of concrete. With the increase of wave steel fiber volume fraction, the workability of fresh concrete was better. The surface area of wave-type steel fiber was significantly lower than that of straight steel fiber, and the slurry volume of steel fiber wrapped in unit volume decreases. The wave steel fiber could improve the fluidity of UHPC to a certain extent.

3.2. Effect of steel fiber content on mechanical properties of UHPC

The compressive strength and flexural strength of UHPC with different volume fraction of steel fiber are shown in the Fig. 2 and Fig. 3. Compressive strength was one of the basic mechanical properties of concrete. UHPC without steel fiber would show a blasting state when it is damaged under pressure, and the incorporation of steel fiber would effectively alleviate the occurrence of blasting phenomenon [16]. The addition of steel fiber could improve the compressive strength of UHPC. When the volume fraction of steel fiber was 0 % ~ 4 %, the compressive strength of UHPC increases with the increase of steel fiber volume fraction, and the compressive strength could reach 163.2 MPa after 28 days. When the volume fraction of steel fiber was more than 4 %, the compressive strength decreased gradually. The friction between steel fiber and matrix increases with the increase of fiber content. The three-dimensional irregular distribution of fiber formed a fiber network skeleton, which could better limit the generation and development of cracks in the foundation. With the continuous increase of the content, the fiber was easy to be unevenly distributed, resulting in the phenomenon of fiber agglomeration [17]. On the contrary, it would reduce the bond strength between the fiber and the matrix, increase the voids in the concrete, and finally lead to the reduction of UHPC compressive strength.

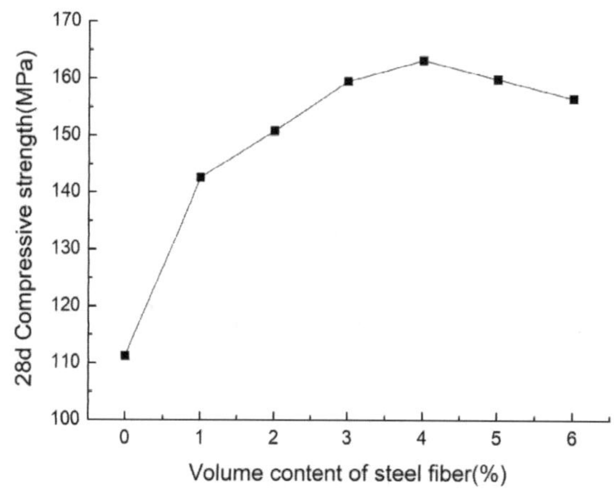

Figure 2. Effect of volume content of steel fiber on UHPC compressive strength

Low flexural strength was one of the main defects of concrete materials. In order to overcome this problem, steel fiber reinforced toughening was a better method. With the increase of steel fiber volume fraction, the flexural strength of UHPC increased first and then decreased. When the volume fraction of steel fiber increased from 0% to 4%, the 28 day flexural strength of UHPC increased gradually, and the maximum value was 45.9 MPa. At this time, when the volume fraction of steel fiber increased by 1%, the average increase of flexural strength was 6.45%. When the volume fraction of steel fiber increased from 4% to 6%, the 28 day flexural strength of UHPC decreased by 9.7%. With the gradual increase of steel fiber content, the steel fiber across the micro cracks played a leading role in strengthening and alleviating the stress concentration at the crack tip. When the fiber content was more than 4%, it was easy to appear agglomeration phenomenon, which made the internal defects of concrete increase, greatly destroyed the compactness of UHPC, and the flexural strength decreased obviously.

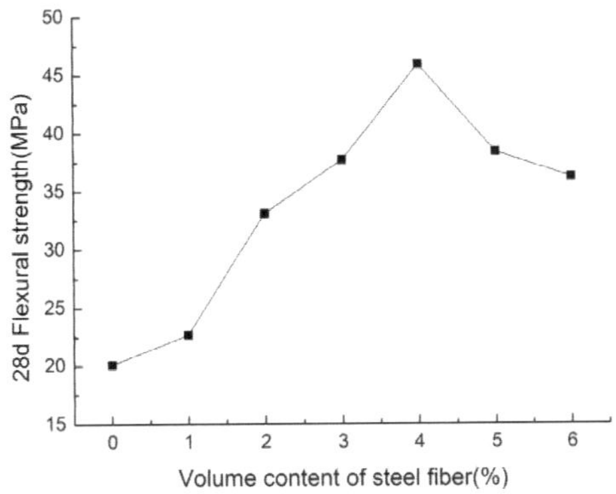

Figure 3. Effect of volume content of steel fiber on UHPC flexural strength

3.3. Effect of steel fiber laspect ratio on mechanical properties of UHPC

Effect of aspect ratio of steel fiber on mechanical properties of UHPC is shown in the figure 4. The 28-day compressive strength of UHPC decreased with the increase of steel fiber aspect ratio under the same volume fraction of steel fiber. When the diameter and content of steel fiber were the same, the number of steel fibers with small aspect ratio was more than that with large aspect ratio. The small aspect ratio of steel fiber dispersed in the matrix could better inhibit the development of micro cracks, resulting in the increase of compressive strength of concrete. When the aspect ratio of steel fiber was 40, the maximum 28 day compressive strength of UHPC was 163.6MPa.

When the volume fraction of steel fiber was constant, the flexural strength of UHPC increased with the increase of the aspect ratio of steel fiber. The steel fiber with larger aspect ratio had longer anchorage length and stronger bonding with the matrix, which could slow down the crack propagation to a large extent. Compared with the steel fiber with small aspect ratio, the steel fiber with large aspect ratio could more effectively inhibit the development of macro cracks, thereby improving the flexural strength of UHPC.

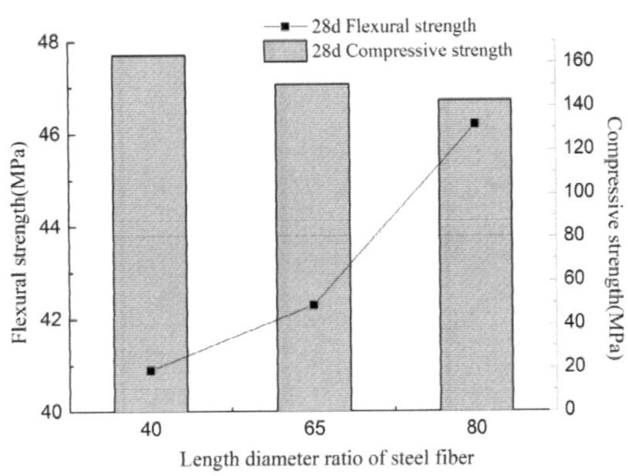

Figure 4. Effects of length diameter ratio of steel fiber on UHPC mechanical properties

3.4. Effect of hybrid steel fiber on mechanical properties of UHPC

Effect of different volume fraction of wave and flat steel fiber on mechanical properties of UHPC is shown in the figure 5. When the volume content of steel fiber was 3 %, the two kinds of steel fibers were mixed in different proportions. With the increase of the volume content of wave-type steel fiber, the 28 - day compressive strength and flexural strength of UHPC showed a gradually increasing trend. With the increase of wave steel fiber content, the mechanical bite force between fiber and cement matrix composite interface increased, and the mechanical properties of UHPC were improved.

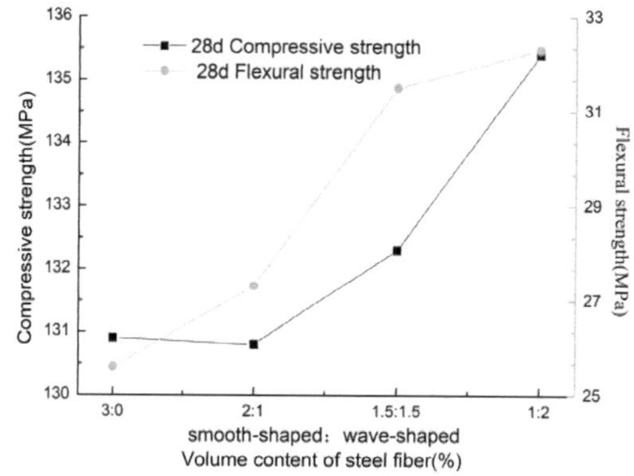

Figure 5. Effect of wave-shaped steel fiber on UHPC mechanical properties

The change of mechanical properties of UHPC prepared by end - hooked steel fiber and straight steel fiber with fiber volume fraction is shown in figure 6. The incorporation of end - hooked steel fiber could effectively enhance the mechanical strength of concrete. The compressive strength of 25 mm end-plated steel fiber was better than that of 30 mm fiber, while the flexural strength of 30 mm end-plated steel fiber was better than that of 25 mm fiber. It was also consistent with the effect of the aspect ratio of steel fiber on the mechanical properties of UHPC studied above.

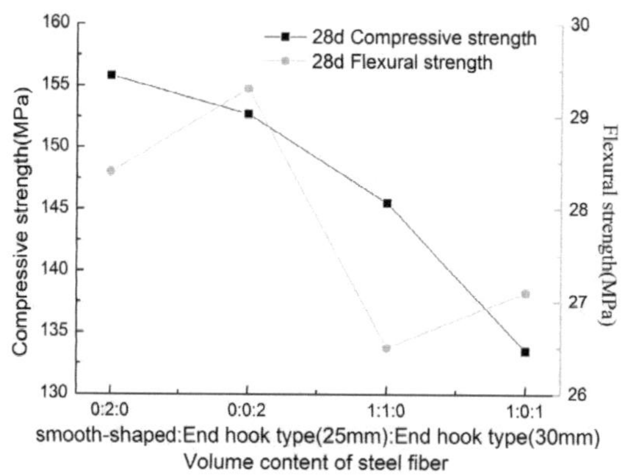

Figure 6. Effect of end hook steel fiber on UHPC mechanical properties

4. Conclusions

In this paper, the effects of steel fiber content, aspect ratio and morphology on the working and mechanical properties of UHPC were studied. The results are as follows:

(1) The incorporation of steel fibers had a certain influence on the working performance of UHPC. With the increase of volume fraction of steel fiber, the slump and expansion of UHPC decreased obviously. With the increase of steel fiber aspect ratio, the slump and expansion of UHPC decrease gradually.

(2) The influence of steel fiber content on the mechanical properties of UHPC was obvious. The compressive strength and flexural strength of UHPC were improved by adding steel fiber properly.

When the steel fiber content reached the turning point of 4%, the mechanical properties of UHPC decreased with the increase of steel fiber content.

(3) When the volume fraction of steel fiber was constant, with the increase of steel fiber length-diameter ratio, the compressive strength of UHPC decreased gradually and the flexural strength increased gradually.

(4) The incorporation of shaped steel fibers would affect the mechanical properties of UHPC. The compressive strength and flexural strength of UHPC increased gradually with the increase of volume fraction of wave type and end hook type steel fiber.

Acknowledgments
The authors want to acknowledge the financial support of the National Natural Science Foundation of China Youth Science Foundation Project (52002041) and China West Construction Group Co., Ltd., science and technology research and development foundation (No. ZJXJ-2019-13) and China State Construction research and development program funding (CSCEC-2019-Z-24).

Conflict of Interests
The authors declare that there is no conflict of interests regarding the publication of this paper.

Data Availability
The raw/processed data required to reproduce these findings cannot be shared at this time due to technical or time limitations.

References
[1] LI Dan 2018 Studies on the influence of stainless steel fiber on the performance of UHPC [D] Hunan University
[2] Nguyen D L, Ryu G S, Koh K T, et al. 2014 Size and geometry dependent tensile behavior of ultra-high-performance fiber-reinforced concrete [J] *Composites Part B: Engineering* **58** 279-292
[3] SHI Cai-jun, XIAO Jiang-fan, CAO Zhang, et al. 2013 Effects of UHPC Constituents on Its Properties [J] *Bulletin of The Chinese Ceramic Society* **32** (6) 1005-1011
[4] Rong Z D, Sun W, Zhang Y S 2010 Dynamic compression behavior of ultra-high performance cement based composites [J] *International Journal of Impact Engineering* **37** (30) 515-520
[5] Feng Naiqian 2009 Development and Application of High Performance Concrete and Ultra-high Performance Concrete [J] *Construction Technology* (04): 5-10
[6] Xu Zhijun *Technology and Application of fiber reinforced concrete* [M] Beijing: China Architecture & Building Press
[7] YAN Chang-wang, JIA Jin-qing, ZHANG Ju 2012 Experimental study of ratio between splitting tensile strength and compressive strength for steel fiber reinforced ultra high strength concrete [J] *Journal of Dalian University of Technology* **52** (2) 233-238
[8] Yoo D Y, Lee J H, Yoon Y S 2013 Effect of fiber content on mechanical and fracture properties of ultra high performance fiber reinforced cementitious composites [J] *Composite Structures* **106** (Complete) 742-753
[9] Yang Zhao-peng 2008 Study on Mechanical Properties of Reactive Powder Concrete [D] Dalian Jiaotong University
[10] Song Yan 2006 The Research on the Tensile Performance of the ultra-high-performance Concrete Reinforced by Grading-Fiber [D] Hunan University
[11] Dong Joo Kim, Antoine E. Naaman, Sherif El-Tawil 2008 Comparative flexural behavior of four fiber reinforced cementitious composites [J] *Cement and Concrete Composites* **30** (10)
[12] Zhidan Rong, Wei Sun 2012 Experimental and numerical investigation on the dynamic tensile behavior of ultra-high performance cement based composites [J] *Construction and Building Materials* **31**

[13] Meng W, Khayat K H 2018 Effect of hybrid fibers on fresh properties, mechanical properties, and autogenous shrinkage of costeffective UHPC [J] *Journal of Materials in Civil Engineering* **30** (4) 1-8
[14] WU Linmei, SHI Caijun, ZHANG Zuhua. et al. 2017 Effects of Steel Fiber on Drying Shrinkage of Ultra High Performance Concrete [J] *Materials Reports* **31** (23) 58-65
[15] Banthia N, Trottier J F 1995 Test Methods for Flexural Toughness Characterisation of Fiber Reinforced Concrete: Some Concerns and Proposition [J] *ACI Materials Journal* **92** (1) 48-57
[16] NIU Xujing 2019 Study on Mechanism of Combined Curing Composed of Precuring in Hot Water and Heating in Dry Air for Optimizing Mechanical Properties and Thermally Explosive Spalling Resistance of Ultra-High Performance Concrete [D] Beijing Jiaotong University
[17] Li Hedong 2009 Experimental Research on Ultra High Toughness Cementitious Composites [D] Dalian University of Technology

Corrosion behavior of Incoloy 840 in molten nitrate salt for CSP plants

Yujun Zhang[1,2], Leilei Xiao[1] and Junhua Liu[1]

[1]Material Science and Engineering School, Xi'an University of Science and Technology, Xi'an 710043, China

[2]Corresponding author's e-mail: zhangyuxiou_dl@163.com

Abstract. As the tube materials of receiver in the solar tower power plant, molten salt corrosion behavior of Incoloy 840 has been studied. In this work， after Incoloy 840 samples were deposed separately in molten Solar Salts (40wt.% KNO_3 + 60wt.% $NaNO_3$) at 560°C for different times (5, 10, 15, 20, 30, 40, 60 and 80h), the weight-loss of samples were measured, the metallurgical structure of as-received Incoloy 840 was observed by optical microscope, the metallurgical structure of corroded films on the samples were observed by SEM, the corrosion products chemistries of the oxide film were analyzed by EDS and XRD. The results showed that the corrosion kinetics of the Incoloy 840 follows typical parabolic rule at 560°C in mixture molten salts (40wt.% KNO_3 + 60wt.% $NaNO_3$), and the corrosion rate of Incoloy 840 is 12.3 μm/year. The corroded films had a bilayer structure, the outer layer consists of Fe_2O_3, and the inner layer consists of γ-(Fe, Ni) and Cr_2O_3.

1. Introduction

As concentrated solar power (CSP) has been commercially operated for 20 years, several technology has been developed on it, Solar tower receiver technology is the quickest one. Solar tower power (STP) plant is designed to focus solar radiations to the receiver on the top of tower in order to achieve higher temperature for heat transfer fluid, which benefit the increase of the Rankine cycle efficiency and the capacity of thermal storage. During the last years, STP rises as a promising solution for providing clean and renewable energy [1-2].

Currently, the molten Solar Salt (40wt.% KNO_3 + 60wt.% $NaNO_3$) is regarded as the most widely used industrial high-temperature heat transfer fluid for STP plants. Because the molting point of Solar Salt is 220°C and the boiling point is 580°C, the temperature of outlet molten salts fluid is raised to 560°C. As a result of this, in STP plant, not only the Rankine cycle efficiency will reach to 40%, but also the thermal storage capacity will significantly increase [3]. As the metal will expose to the molten salts, with an increase in temperature, more serious safety problem rises up, such as crack failure of the receiver tube due to the high-temperature corrosion. So in that case, the original 316 stainless-steel receiver tubes were suggested to replace by advanced high nickel alloy ones in the Solar Two STP project [4].

In order to keep the molten salts STP plant safe, the high-temperature molten Solar Salt corrosion behavior of alternative receiver tube metal materials needs to be comprehend well [5-6]. Because of their good anti-corrosion properties in molten salts and good mechanical properties in high temperature, high chromium-nickel steels have been considered as optional materials. However, there are too many types of heat-resisting high chromium-nickel steels to choose from. In fact, the lack of comprehension on corrosion behavior imposed restrictions on the alloy selection for exposed to molten Solar Salt.

There are some studies on high-temperature corrosion performance of high chromium-nickel steels, 316 stainless-steel and Incoloy 800, in mixed molten Solar Salt [7-8], but the high-temperature anti-corrosion property of 316 stainless-steel is not gratifying, and the price of Incoloy 800 is rather stiff. Incoloy 840 has better corrosion performance than 316 stainless-steel in mixed molten nitrate salts at high temperatures, and it costs less than Incoloy 800 because of its low nickel content. There is still no systematic study on corrosion performance of Incoloy 840 in mixed molten nitrate salts at high temperatures [9-11]. The high-temperature anti-corrosion property of Incoloy 840 has not be systematic researched.

In the paper, as the tube materials of receiver in the STP plant, molten salt corrosion behavior of Incoloy 840 has been studied in several ways. After Incoloy 840 samples were submerged separately in molten Solar Salt at 560°C for different times (5, 10, 15, 20, 30, 40, 60, and 80h), the weight-loss of samples were measured, the metallurgical structure of as-received Incoloy 840 was observed by optical microscope, the metallurgical structure of corroded films on the samples were observed by SEM, the corrosion products chemistries of the oxide film were analyzed by EDS and XRD.

In this word, after Incoloy 840 samples were deposed separately in molten Solar Salts (40wt.% KNO3 + 60wt.% NaNO3) at 560°C for different times (5, 10, 15, 20, 30, 40, 60 and 80h), the weight-loss of corrosion samples were measured, the morphology of as-received Incoloy 840 was observed by optical microscope, the morphology of corroded films on the samples were observed by SEM, the corrosion products chemistries of the oxide film were analyzed by EDS and XRD.

2. Experimental procedures

2.1. Preparation of samples for corrosion study

The chemistry of as-received Incoloy 840 was given by manufacturers and list in table 1. Corrosion samples with size of 10×10×2 mm were cut from as-received Incoloy 840 plates. Then the samples were washed out, grinded with 1000-grit SiC abrasive paper, ultrasonically cleaned for 5 minutes in ethanol and acetone in turn, and dried up in air. Final, the initial weights of the Incoloy 840 samples were measured by an analytical balance with a resolution of 0.1 mg (JY600).

Table 1. Nominal composition of as-received Incoloy 840 [wt.%]

Ni	Cr	Fe	Si	Mn	Ti	C
18	20	balanced	1	1	1	<0.05

2.2. Preparation of Solar Salt mixture

Commercial potassium and sodium nitrates were purchased from LIANDA Chemical, with a composition according to the manufacturer summarized in Table 2. In order to prepare the molten Solar Salt eutectic mixture, the nitrates were dried up in oven at 120°C, then mixed uniformity in the right proportion and heat to melt in chamber furnace, and keep at a uniform temperature about 560°C.

Table 2. Nitrates salt composition in weight [wt.%]

	Purity	Chloride	Sulfate	Carbonate	Insolubles	Moisture
KNO_3	>99.7	<0.01	<0.005	<0.01	<0.25	<0.1
$NaNO_3$	>99.9	<0.01	<0.01	<0.01	<0.01	<0.02

2.3. Corrosion in the molten Solar Salt

In order to test the high-temperature molten Salt-induced corrosion behavior of Incoloy 840, the following steps had been adopted. Firstly, the metal samples were submerged respectively in the alumina crucibles with the enough amount of molten solar salt to submerge the whole sample surface

IOP Conf. Series: Materials Science and Engineering 1167 (2021) 012002 doi:10.1088/1757-899X/1167/1/012002

for different times (5, 10, 15, 20, 30, 40, 60 and 80h). Secondly, three parallel corroded samples were extracted from the alumina crucible with molten Solar Salt at the right slot, then cooled in air and rinsed with distilled water. Thirdly, the corroded samples were immersed in 10% nitric acid. Finally, the corroded samples were weighed by an analytical balance after being cleaned and dried.

2.4. Samples characterization

The metallurgical structure of as-received Incoloy 840 was observed by OM (MH-5L). The metallurgical structure of corroded films on the samples were observed by SEM (JSM-6390A), The corrosion products chemistries of the corrode film were analyzed by EDS and XRD (XRD-7000).

3. Results and discussion

Figure 1 shows the Optical metallurgical structure of the initial Incoloy 840 alloy. The Incoloy 840 alloy has the equiaxed grains with average size of 100±20μm. From the point of view of physical metallurgy, Incoloy 840 alloy has a single-phase FCC austenitic structure.

Figure 2 shows the weight-loss curves of corroded samples in the Solar Salt at 560°C. As shown in figure 2, the weight-loss of the alloy increases dramatically in the initial 30h, and then the increasing rate slow down. Accordingly, the weight-loss increased in a declining rate in particular after 30h. The results of weight-loss experiment show that the corrosion kinetics of the Incoloy 840 alloy follows typical parabolic rule when exposed in molten Solar Salt at 560°C. According to this rule, the rate of the corrosion reaction is inversely proportional to the thickness of corrode film or the weight of oxide formed. Based on the data of figure 2, the corrosion rate of Incoloy 840 in molten Solar Salt is calculated from equation (1) and (2), and the result is 12.3μm/year.

$$\text{Corrosion Rate(mm/year)} = \frac{\Delta M}{\rho S T} \tag{1}$$

$$\Delta M^2(\text{g}) = \text{K}T \tag{2}$$

where: ΔM=Weight loss (g), K = parabolic rate constant, T = corrosion time(h), S = surface area (cm^2), ρ= density = 8.0g/cm^3(for Incoloy 840 alloy).

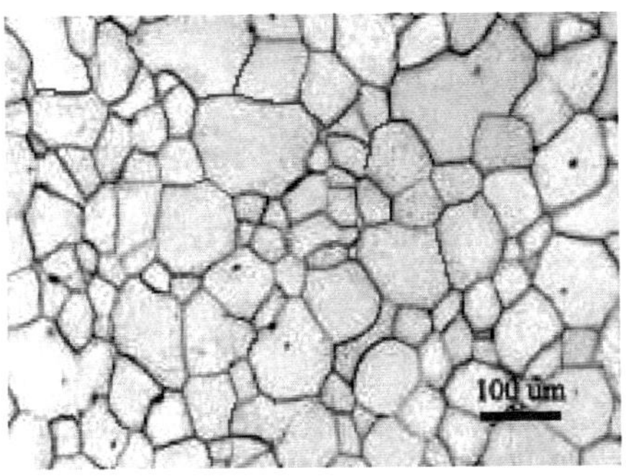

Figure 1. Optical photomicrograph of the as-received Incoloy 840.

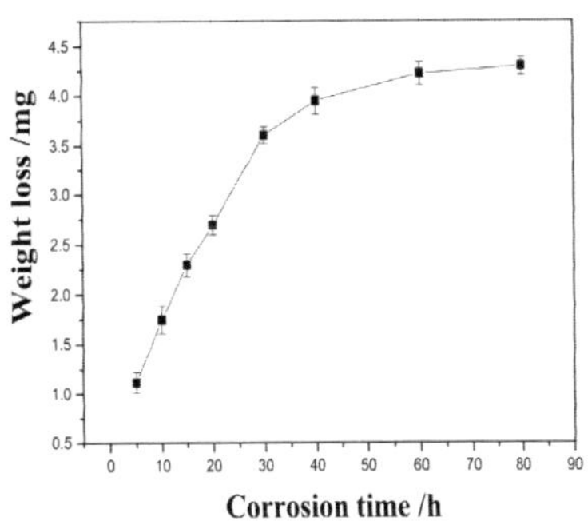

Figure 2. Weight-loss curve of the Incoloy 840 in molten Solar Salt at 560°C.

Figure 3 shows the XRD pattern of the corroded surface of Incoloy 840 alloy samples which had exposed to the molten Solar Salts for 80h at 560°C. The major phases compositions in the oxide film are Cr_2O_3 and Fe_2O_3, maybe also $NiCr_2O_4$.

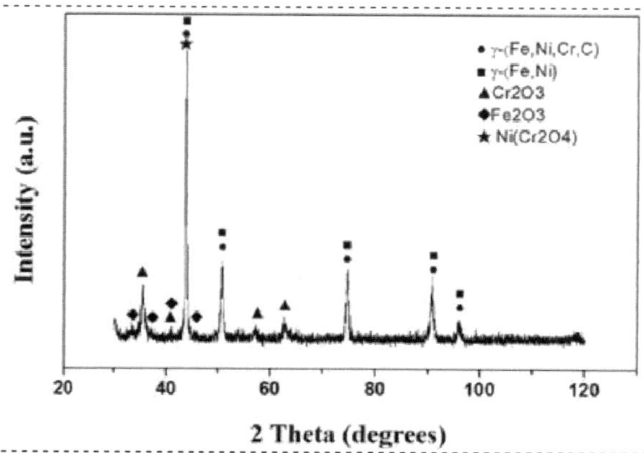

Figure 3. XRD pattern of the corroded surface of Incoloy 840 sample exposed to the molten Solar Salts for 80h at 560°C.

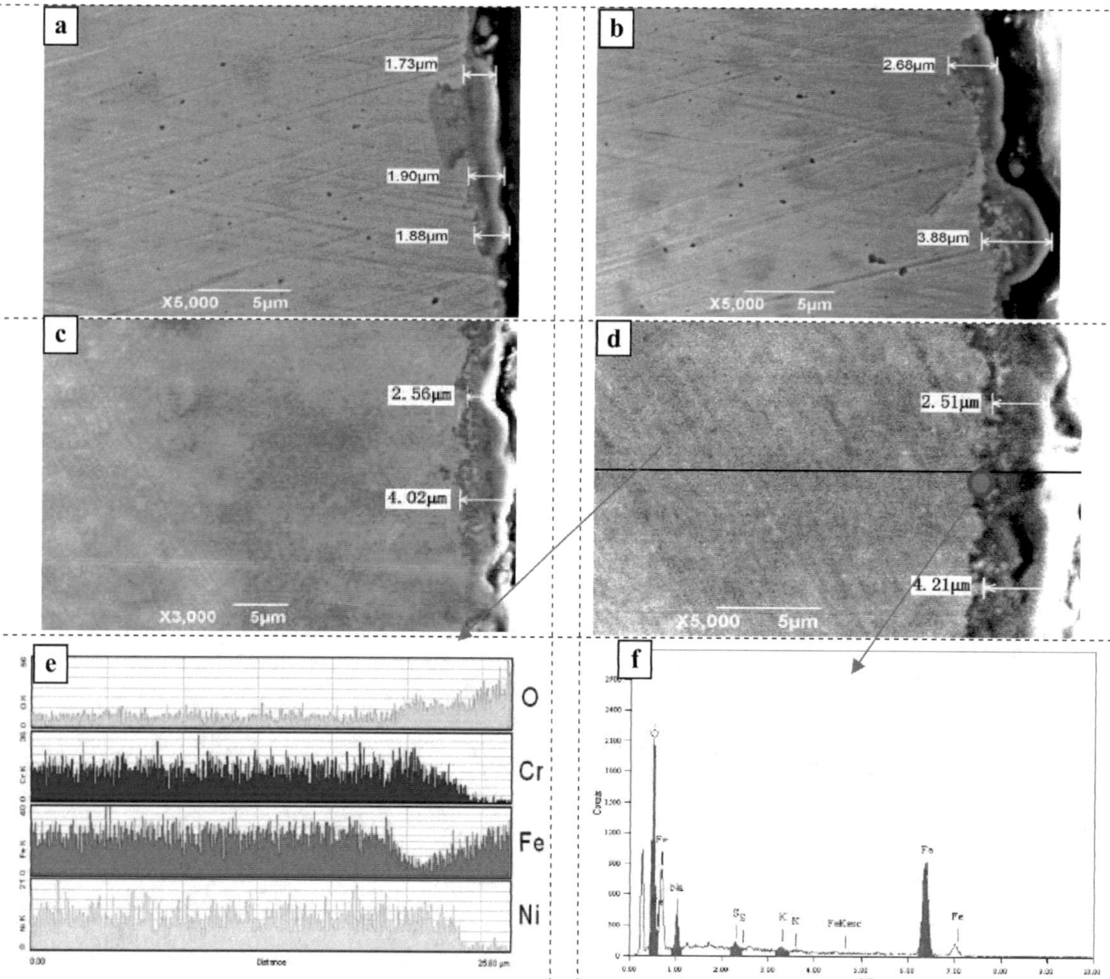

Figure 4. SEM morphology of the cross-section of the corroded samples, exposed to the molten Solar Salts for 15h(a), 30h(b), 60h(c) and 80h(d) at 560°C, as well as (e) demonstrates the linear element distributions from inside to the surface detected by EDS, and (f) EDS pattern of the outer layer of oxide film.

Figure 4 shows the SEM metallurgical structures and EDS patterns of the cross-section of corrosion Incoloy 840 samples. As it shown in figure 4(a), while the Incoloy 840 exposed to the molten Solar Salts for 15h at 560°C, the thickness of corrode film increased to about 1.5μm, and then increased to about 2.5μm for 30h, as shown in figure 4(b), and 3.5μm for 60h, as shown in figure 4(c), at last 4.2μm for 80h in figure 4(d). The thickness of corrode film increases with corrosion time, and it was well matched with the weight-loss experimental results.

Figure 4(e) shows the EDS patterns of the corrode film. As shown in the figure 4(e), the corrode film composition contains only two elements: Fe and O, it illustrated that the outer layer of the corrode film is composed of iron oxide. The result was well matched with the XRD test result, as figure 3 shown. Figure 4(f) demonstrates the linear element distributions from inside to the surface. The corrode films shows a bilayer structure, the outer layer consistes of O and Fe elements, however, the inner layer includes Fe, O, Ni and Cr elements. It illustrated that the outer layer of the corrode film is composed of iron oxide and inner layer consisted of iron oxide, chromium oxide, nickel oxide or Fe-Cr complex oxides. In connection with figure 3, the oxide films had a bilayer structure: the outer layer of the corrode film is Fe_2O_3 and the inner layer consisted of γ-(Fe,Ni) and Cr_2O_3.

4. Conclusions

In this paper, the molten salt (40wt.% KNO_3 + 60wt.% $NaNO_3$) corrosion behavior at 560°C of Incoloy 840, as the tube materials of receiver in the STP plant, has been studied in several ways.

The results of weight-loss experiment show that the corrosion kinetics of the Incoloy 840 alloy follows typical parabolic rule when exposed in molten Solar Salt at 560°C, and the corrosion rate is 12.3μm/year. Goods et al. [12] analysed corrosion behaviour of 316 stainless steels in-depth and the corrosion rates is 15 μm/year at 570°C. The analysis of the corrosion products chemistries of the oxide film shows the oxide films had a bilayer structure, the outer layer consists of Fe_2O_3, and the inner layer consists of γ-(Fe, Ni) and Cr_2O_3.

References

[1] Ding W J, Gomez-Vidalb J, Bonka A and Bauerc T 2019 *Sol. Energy Mater Sol. Cells* **199** 8-15
[2] Reddy V S, Kaushik S C, Ranjan K R and Tyagi S K 2013 *Renew. Sust. Energ. Rev.* **27** 258
[3] Sun H, Su X Z, Zhang P and Wang J Q 2017 *Corros Sci Prot Technol* **29** 282-290
[4] Pillai R, Raiman SS and Pint B A 2021 *J. Nucl. Mater.* **546** 1-13
[5] Yang Z and Garimella S V 2013 *Appl Energ* **103** 256
[6] Kuravi S, Trahan J, Goswami D Y, Rahman M M and Stefanakos E K 2013 *Prog. Energy Combust. Sci.* **39** 285
[7] Hu H X, Zheng Y G and Qin C P 2010 *Nucl. Eng. Des.* **240** 2721-2730
[8] Ahmed O, Zhou L, Mohajeri N and Sohn Y H 2014 *Adv. Mater. Res.* **922** 13-17
[9] Tzvetkoff T and Gencheva P 2003 *Mater Chem Phys* **83** 897
[10] Zahrani E M and Alfantazi A M 2012 *Corros Sci* **65** 340
[11] Liu M, Bell S, Segarra M, Steven Tay N H, Will G, Saman W and Bruno F 2017 *Sol. Energy Mater. Sol. Cells* **170** 1-7
[12] Goods S H, Bradshaw R W 2004 *J. Mater. Eng. Perform.* **13** 78-87

Determination of glass transition temperature and application for fabrication of polyphosphazene with high performance

Shaojun Wu[1], Xiang Xue[1,2]

[1]School of Materials Science and Engineering, Harbin Institute of Technology, Harbin 150001, China

[2]Corresponding Author's E-mail: xxue@hit.edu.cn

Abstract. To determine the relations of performance and structure accurately, polyphosphazenes with various substitution rates of phenoxy and ethoxy were prepared by nucleophillic substitution reaction, their structures and glass transition temperature (Tg) were characterized by NMR and DSC, respectively, and theoretical Tg of polyphosphazenes containing various structural units were calculated by Fox equation. We can find that Tg of polyphosphazenes decreased with increase of ethoxy contents, which improved low temperature resistance of the polymer. According to the measured value and calculated value of Tg, as well as the deduced mathematical expression of Tailor-Gorden equation, the relationships between Tg and ethoxy contents were proposed, which demonstrated that a plot of the Tg vs. ethoxy contents yielded straight lines when polymer chain owned higher substitution rates of phenoxy or ethoxy, while Tg of polymers with more structural units of phenoxy and ethoxy would be adjusted by a parameter of ω_B. Based on the proposed relation, the application of fabrication on polyaryloxyphosphazene elastomer with better mechanical properties and low temperature resistance was successfully carried out. Therefore, the proposed methods can be useful in the prediction of polymer Tg and can also be used for structure prediction of polymers to match a target Tg value by this means, which are likely to have broad applications in polymer science fields.

1. Introduction

The polyphosphazenes comprise of an extensive family of polymers built on a -$[P=N]_n$- backbone owing to high reactivity which can be linked to wide range of side chains, leading to the products with better heat resistance, flame retardance, antioxidant, solvent resistance and other excellent physical and chemical properties as compared to organic polymer materials with carbon chains [1-6]. The properties of better film-forming, resistance to high temperature and erosion can be obtained by poly (aryloxyphosphazenes), but their higher Tg of over 0℃ result in a narrow work temperature range, which highly restrict the applications of poly (aryloxyphosphazenes) [7,8]. The side groups have a certain influence on the Tg of the polymer, thereby we envision that Tg of poly (aryloxyphosphazenes) can be appropriately reduced by introducing some alkoxy into poly (aryloxyphosphazenes) and controlling their proportions [9,10], which can partly retain better properties of poly (aryloxyphosphazenes). However, a polyphosphazene with specific Tg and structural characteristic, containing phenoxy and ethoxy, could not be easily prepared owing to highly uncertain of theoretical usage amount of phenoxy and ethoxy groups during synthesis.

In order to determine the relationships of side groups with Tg of polymers, we prepared a series of

IOP Conf. Series: Materials Science and Engineering 1167 (2021) 012003 doi:10.1088/1757-899X/1167/1/012003

polyphosphazenes containing different proportions of phenoxy and ethoxy by nucleophilic substitution reaction, the prepared polymers consist of three kinds of structural units, which are a phosphorus link with two phenoxy (unit A), a phosphorus link with a phenoxy and an ethoxy (unit B) as well as a phosphorus link with two ethoxy (unit C). Then, the measured values of Tg by DSC and calculated values of Tg by Fox equation of polyphosphazenes with different substitution rates were obtained, and the approximate relationship between theoretical calculated values of Tg and contents of ethoxy were fitted, we found that the measured values were always near to or on the fitting line, which was in agreement with derived equation from Tailor-Gorden equation and indicated the truth of obtained relation.

On the basis of the combination of theoretics and experiments methods mentioned above, the deduced relations of Tg and structure of polyphosphazenes were confirmed and could accurately predict Tg by structure or design a specific structure by needed Tg, which could better guide the synthesis of the polyaryloxyphosphazene elastomer with better mechanical properties and low temperature resistance and is likely to have broad applications for structure design of other high performance polymers with structure units.

2. Experimental Details

2.1. Materials
Tetrahydrofuran (THF), petroleum ether, phenol, xylene, ethanol, n-heptane and trichlorobenzene were purchased from Beijing Chemical Works, China, and all of the chemical reagents were of analytical grade. Sodium and hexachlorocyclotriphosphazene (HCCP) recrystallized with n-hexane were commercially purchased. All other additives were commercially purchased and used as received.

2.2. Synthetic procedures
Polydichlorophosphazene (PDCP) was synthesized as a modified way in our lab [11, 12], sodium salt of ethanol and phenol were prepared in the flask equipped with ice-water bath by the reaction of ethanol and phenol with sodium for several hours, respectively. For polyphosphazenes with higher substitution rates of phenoxy, certain amount of PDCP in xylene was dropped into sodium salt of ethanol solution via an addition funnel. The reaction mixtures were stirred at 0℃ under N_2 for 1~2 hrs firstly, and then sodium salt of phenol was dropped into the mixtures at room temperature under N_2 for 12 hrs, subsequently the mixtures were heated up to 110℃ in an oil bath equipped with reflux condenser and stirred for 24 hrs. While for polyphosphazenes with lower contents of phenoxy, certain amount of PDCP in THF was dropped into sodium salt of ethanol solution at 0℃ under N_2 for 12 hrs firstly, then sodium salt of phenol was added into mixtures and the reaction was carried out at 60℃ for 48 hrs. After the reactions were finished, reaction mixtures were precipitated and dried.

2.3. Characterization
^{31}P-NMR and ^{1}H-NMR analyses were performed on a Bruker AV- III instrument operated at 400 MHz, the concentration of specimens which were dissolved in $CDCl_3$ was 10 mg/mL. The Tg values of polymers were measured using a TA Instrument DSC 2920 calibrated with an indium standard. Samples (5~10 mg) were equilibrated against ambient temperature and humidity prior to measurement. Measurements were made under a nitrogen atmosphere at 10℃ /min, samples were cooled from room temperature to -120 or -60℃ as required and then heated in a second scan to 25℃. The glass transition temperature was determined from the second heating scan by the half Cp extrapolated tangent method.

3. Results and discussion

3.1. Synthesis and glass transition temperature of polymers
The detailed analysis of the characterization data of all the as-prepared eight polymers (polymer 1-polymer 8) has been summarized in Table 1. The phenoxy and ethoxy side groups percentages were

determined by the ^{1}H-NMR and ^{31}P-NMR, which were fitted by XPS Peak Fit software to calculate the ratios of three kinds of structural units accurately. The ^{31}P-NMR spectra of polymer1-polymer8 (Figure 1) showed the peaks at near δ= -19.83, -14.40 and -8.87 ppm corresponding to the structure of unit A, unit B and unit C, respectively, the peak of δ=-14.40 ppm presented enhancement and then disappearance with increase of the substitution rate of ethoxy. The ^{1}H-NMR spectra of polymer 1-polymer 8 were exhibited in Figure 3, and the peaks at near δ=1.09, 3.80 and 7.05 ppm belonged to protons of methyl, methylene and benzene ring, respectively. The five hydrogens of benzene ring showed a peak at δ=7.05 ppm owing to same chemical environment and disappeared with increase of the contents of ethoxy. The ^{1}H-NMR analyses were compatible with structures of polymer 1 -polymer 8 recorded by ^{31}P-NMR, which indicated that polymers have been successfully prepared according to our design.

Table 1. ^{31}P-NMR, ^{1}H-NMR and Tg analysis of polymer 1~polymer 8

	^{31}P-NMR		^{1}H-NMR		Average substitution rate		Tg/℃
	phenoxy	ethoxy	phenoxy	ethoxy	phenoxy	ethoxy	
polymer1	100%	0	100%	0	100%	0	8.2
polymer2	88.7%	11.3%	89.8%	10.2%	89.3%	10.7%	-5.5
polymer3	65.5%	34.5%	66.4%	33.6%	66.0%	34.0%	-25.6
polymer4	48.1%	51.9%	50.5%	49.5%	49.3%	50.7%	-26.7
polymer5	23.5%	76.5%	23.4%	76.6%	23.5%	76.5%	-61.3
polymer6	19.4%	80.6%	18.3%	81.7%	18.9%	81.1%	-63.4
polymer7	8.8%	91.2%	5.2%	94.8%	7.0%	93.0%	-78.5
polymer8	0	100%	0	100%	0	100%	-86.6

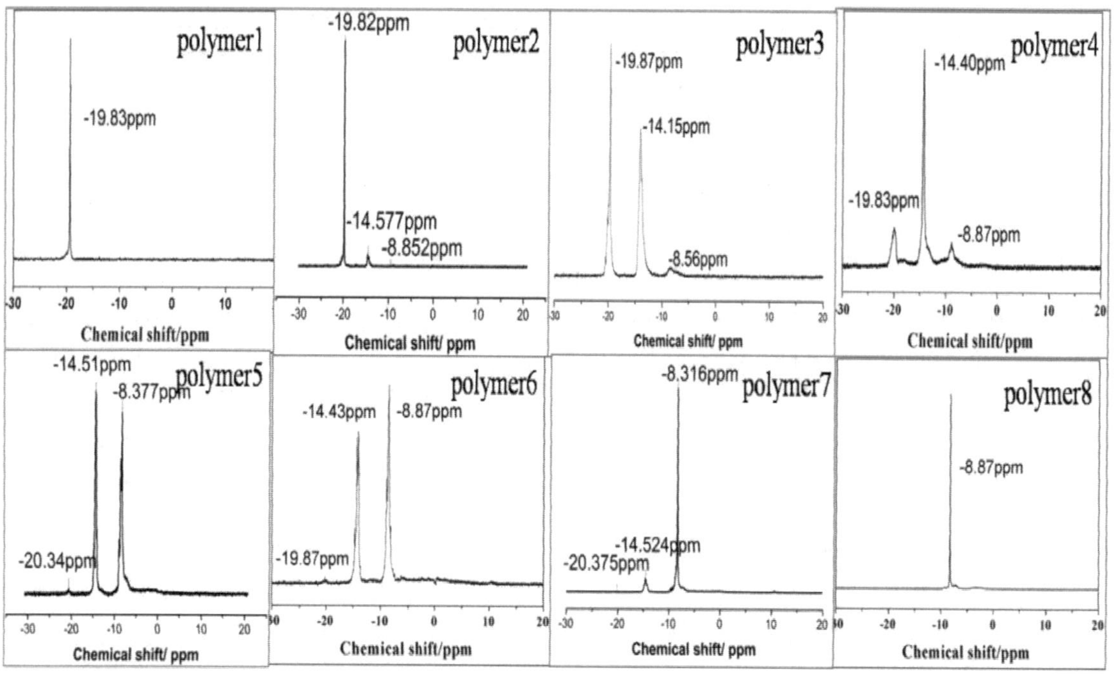

Figure 1. ^{31}P-NMR spectra of polymer 1~polymer 8.

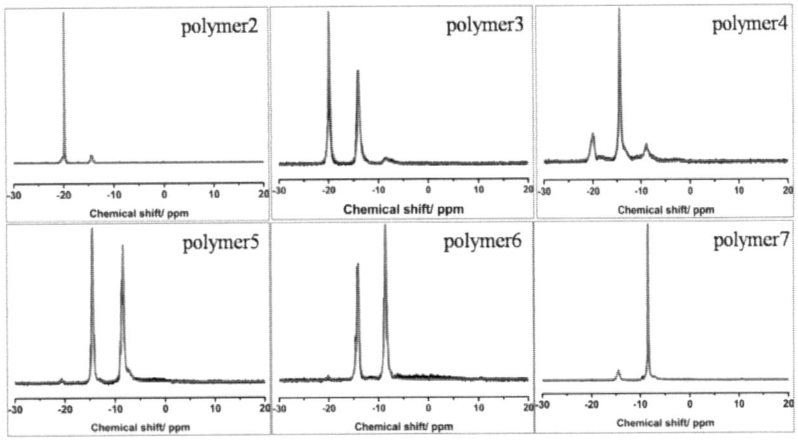

Figure 2. the fitted ^{31}P-NMR spectra of polymer 2~polymer 7 by XPS peak software.

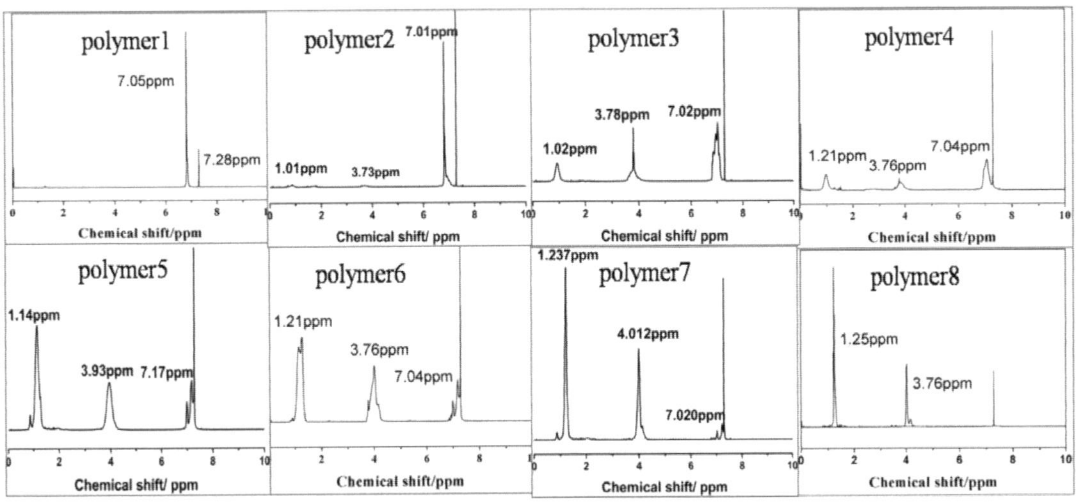

Figure 3. ^{1}H-NMR spectra of polymer 1~polymer 8

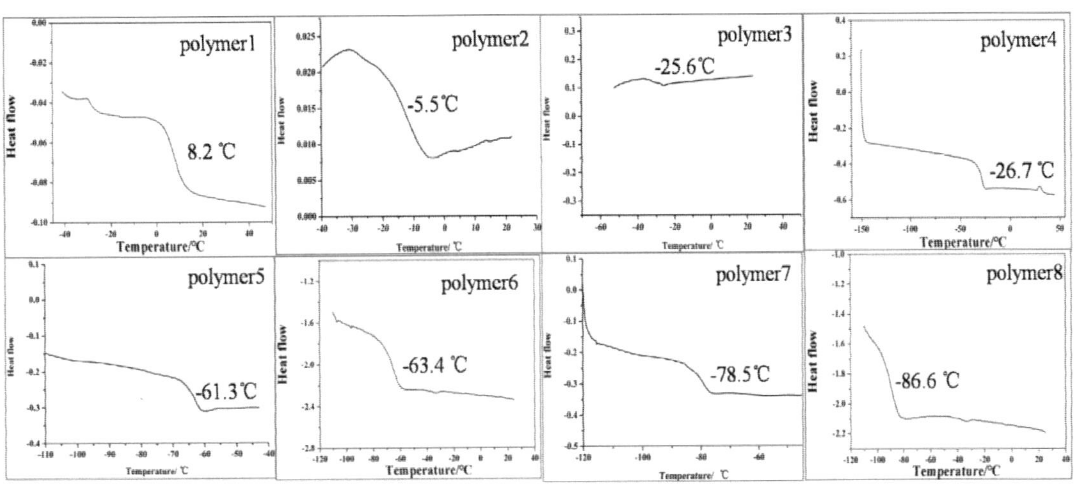

Figure 4. DSC curves of polymer 1~polymer 8

The Tg curves of polyphosphazenes with diverse substitution rates of ethoxy were depicted in Figure 4, which were measured by DSC under a nitrogen atmosphere. According to the values of them shown in Table 1, the Tg of polyphosphazenes gradually decreased with increasing the substitution rate of ethoxy, the reason is that the introduction of certain amount of ethoxy destroys the regularity of the poly(phenoxyphosphazene) chain, which is not conducive to the regular arrangement of the chain. These result in the enhancement of free volume, leading to the improvement of chain segment's mobility, on the other hand, the steric hindrance of ethoxy in the polymer chains is smaller as compared to the phenoxy, indicating the hindered intramolecular rotation could be effectively reduced and increasement of flexibility.

3.2. Effect of ethoxy substitution rate on Tg and structural units

Tg of polyphosphazene only containing unit B can not be measured directly by DSC because the polyphosphazene is difficult to be synthesized in terms of synthetic method, nevertheless, polyphosphazenes with three types of randomly distributed structural units can be prepared and measured by DSC, meanwhile, the Tg of poly[dis-aryloxy]phosphazenes and poly[dis-ethoxy]phosphazenes could be measured, i.e., Tg_A and Tg_C can be obtained, then according to the relationships of structural units' proportions from ^{31}P-NMR with Fox equation:

$$\frac{1}{Tg}=\frac{\omega_A}{Tg_A}+\frac{\omega_B}{Tg_B}+\frac{\omega_C}{Tg_C}$$

Tg_B of polyphosphazene merely containing unit B can be calculated. In order to achieve an accurate value of Tg_B, six values of Tg_B from polymer2~polymer7 and an average value of Tg_B=-22.05 ℃ were obtained through the above method, the consequences of them were summarized in Table 2.

Table 2. Tg of structural unit B

Tg_{B1}	Tg_{B2}	Tg_{B3}	Tg_{B4}	Tg_{B5}	Tg_{B6}	Tg_B
-48.22℃	-42.8℃	-16.01℃	-33.55℃	-27.22℃	-42.9℃	-35.12℃

On the basis of the arguments presented above, when we are aware of Tg_A=8.2 ℃, Tg_B=-35.12 ℃, Tg_C=-86.6 ℃, $\omega_A=1-\omega_C-\omega_B$, $\omega_{ethoxy}=\omega_C+\omega_B/2$ and Tailor-Gorden equation of terpolymer:

$$Tg=\frac{\omega_A T_{gA}+k_1\omega_B T_{gB}+k_2\omega_C T_{gC}}{\omega_A+k_1\omega_B+k_2\omega_C}$$

where Tg is glass transition temperature of terpolymer; Tg_A, Tg_B, Tg_C, ω_A, ω_B and ω_C are glass transition temperature of structural unit A, unit B, unit C and their proportions in polymer chain, respectively; $k_1=\alpha_B/\alpha_A$, $k_2=\alpha_C/\alpha_A$, α_A, α_B and α_C are coefficients of thermal expansion of three kinds of structural units, and generally we can regard k_1 and k_2 as 1, a derived equation from Tailor-Gorden equation could be deduced:

$$Tg\,(℃)=-94.8\omega_{ethoxy}+8.2+4.08\omega_B \tag{1}$$

It is apparent that the correlation of Tg with ω_{ethoxy} approximately is linear if the value of ω_B is samller, which corresponding to larger amount of phenoxy or ethoxy in polymer chains from ^{31}P-NMR. Therefore, a linear fit of the plot of the theoretical calculated polymer's Tg (in ℃) vs. ethoxy substitution rates can be gained in Figure 5 after obtaining calculated Tg from parameters of Tg_A, Tg_B, Tg_C, ω_A, ω_B, ω_C and Fox equation, and Figure 5 demonstrated that measured values of Tg were near by the theoretical calculated values on the fitted line, suggesting that the proposed method can be used as a Tg prediction tool. Additionally, the measured values of Tg were approximately equal to the theoretical calculated values under the circumstance of more contents of phenoxy or ethoxy in polymer chain, while the substitution rates of ethoxy and phenoxy were 50%, respectively, there is a largest error of less than 10 K between measured values and theoretical calculated values, these must be that more randomly distributed ways of these two side groups in polymer chain could be achieved, resulting in the arrangement ways of chain segments maybe are distinct from a polymer owing same ethoxy substitution rate of 50%. These results are consistent with the deduced equation from Tailor-Gorden equation, that is, when the proportions of a certain side group in polymer chain is higher, Tg (℃)≈-94.8ω_{ethoxy}+8.2, which is almost in accordance with the fitted equation of Tg (℃) =-92.35ω_{ethoxy}+9 (R^2=0.9326). While the contents of both phenoxy and ethoxy are higher, 4.08ω_B have a larger impact on Tg of polyphosphazenes, leading to a larger error between measured values and theoretical calculated values.

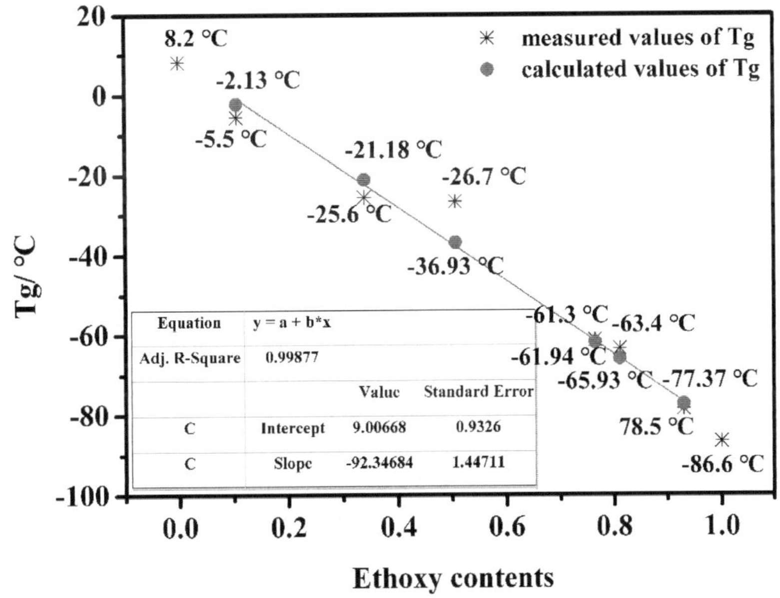

Figure 5. Relationship between Tg and ethoxy contents

Considering the above analysis, when the substitution rates of a certain side group are higher, the target Tg and proportions of structural units of polymer can be achieved by following relationships under the known value of ω_{ethoxy}:

$$1/Tg\,(K) = \omega_A/281.35 + (1-\omega_A-\omega_C)/238.03 + \omega_C/186.55 \quad (2)$$

$$\omega_{ethoxy} = \omega_C + (1-\omega_A-\omega_C)/2 \quad (3)$$

$$Tg\,(℃) = -92.35\omega_{ethoxy} + 9 \quad (R^2=0.\ 0.9326) \quad (4)$$

while the contents of both phenoxy and ethoxy are higher, 4.08ω_B have a larger impact on Tg of polyphosphazenes. The values of ω_B of polymer 3 ~ polymer 6 are higher from ^{31}P-NMR, which range from 0 to 0.55. Therefore, a correlation between the polymer's Tg, ω_{ethoxy} and ω_B, as expressed in Eq. (5):

$$Tg\,(℃) = -94.8\omega_{ethoxy} + 8.2 + 4.08\omega_B,\ \omega_B \in (0,\ 0.55] \quad (5)$$

and then a range of Tg of polyphosphazene can be determined by it.

3.3. Verification of the obtained relationships

To further objectively confirm the validity of our approach as a basis for a Tg prediction method, a polyphosphazene with target Tg=-50 ℃ was determined to be prepared through the above method.

The value of ω_{ethoxy} is 0.64 by calculating from equation (4), while a range values of 0.614 ~ 0.638 were obtained from the deduced equation (5). Thereby we preliminarily set ω_{ethoxy} as 0.64 and the polyphosphazene was synthesized by above mentioned way. ^{31}P-NMR and ^{1}H-NMR of the polyphosphazene were shown in Figure 6, and our designed product have been prepared because the substitution rate of ethoxy is 0.638. Figure 7 indicated that Tg of the polyphosphazene is -48.34 ℃, which is approximately equal to the values obtaining from equation (4) (-49.91 ℃) and equation (5) (-50.03 ~ -52.28 ℃). Hence, the relationships we obtained are more effective and precise approach to be used for structure prediction of polymers to match a target Tg value.

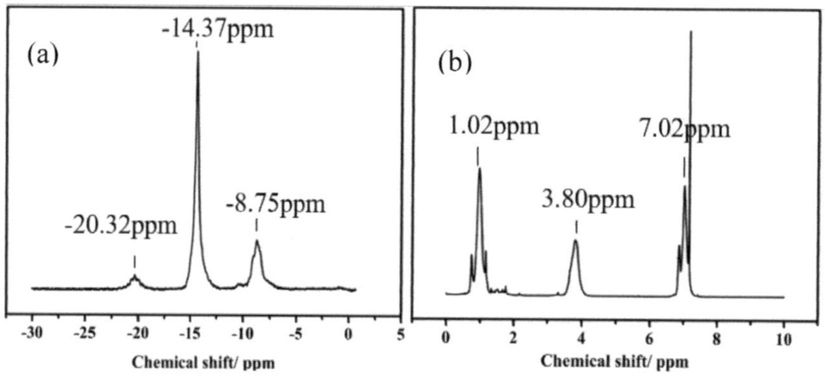

Figure 6. (a) ^{31}P-NMR and (b) ^{1}H-NMR spectra of polymer with the ethoxy substitution rate of 0.638

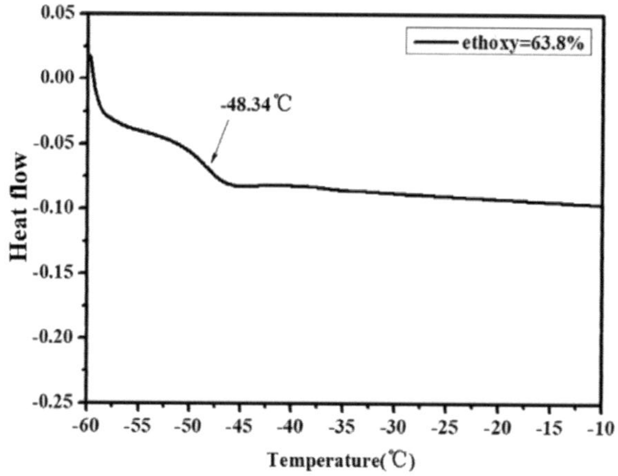

Figure 7. DSC curves of polymer with the ethoxy substitution rate of 0.638

3.4. Preparation of polyaryloxyphosphazene with low temperature resistance

Generally, polyaryloxyphosphazenes with higher glass transition temperature, which is over 0 ℃, can not be used as elastomer when they are under very cold condition. Therefore, these kind of elastomers with poor low temperature resistance must be modified to reduce their Tg according to appropriate method. Giving the above study, polyphosphazene with Tg of -50 ℃ can be prepared by introducing ethoxy of 64% into polyphosphazene containing phenoxy. Herein, polyaryloxyphosphazene with low temperature resistance was synthesized by regarding p-ethylphenoxy and eugenol as phenoxy and ethoxy, respectively. The contents of side groups in chain were listed in the Table 3, and the synthesized product was characterized by NMR and DSC as well. As shown in DSC curve of product, the Tg of polyaryloxyphosphazene is approximately -46.29 ℃, which do not exhibit a significant difference with target value.

IOP Conf. Series: Materials Science and Engineering 1167 (2021) 012003 doi:10.1088/1757-899X/1167/1/012003

Table 3. Proportions of side groups of polyaryloxyphosphazene

side groups	ethoxy	$HORCH=CH_2$	phenoxy	p-ethylphenoxy
theoretical proportion	60%	5%	30%	5%
actual proportion	59.75%	5.76%	30.66%	3.80%

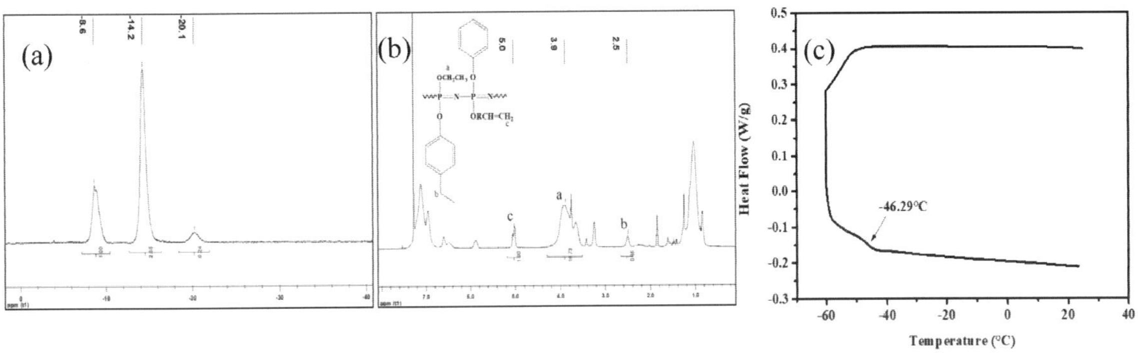

Figure 8. (a) ^{31}P-NMR, (b) ^{1}H-NMR and (c) DSC curves of synthetic product

The as-synthesized polyaryloxyphosphazene with excellent low temperature resistance was vulcanized under 160 ℃. It is apparent that the ability of vulcanization of polyaryloxyphosphazene is good according to vulcanization curve in Figure 9. Standard specimens for tensile testing were hot pressed at a plate vulcanizing machine under 160 °C and 15 MPa for 38 min. To investigate the mechanical properties, six samples were fabricated by same process, and the mechanical properties of three representative samples after vulcanizing were presented in Figure 10, the average tensile strength of the sample is maintained at around 5 MPa, and elongation at break could even reach more than 300%. The SEM picture shows that the sample exhibits the characteristic of ductile fracture, contributing to higher elongation. The mechanical properties with error range of polyaryloxyphosphazene elastomer were depicted in Table 4, meaning that polymer with better mechanical properties and low temperature resistance was obtained after modification by above approach.

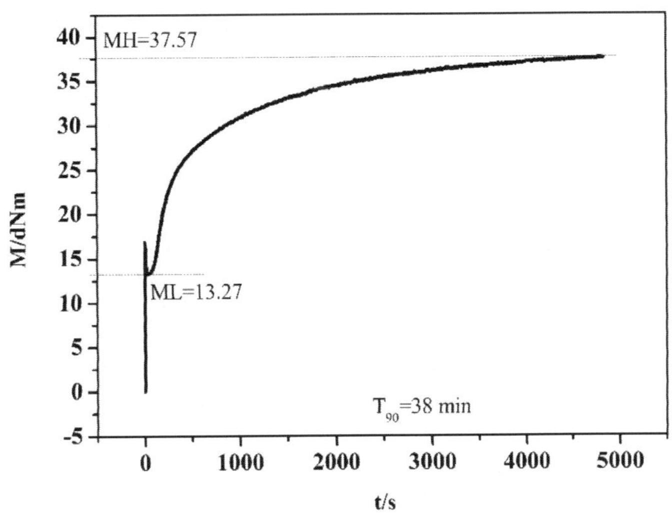

Figure 9. Vulcanization curve of polyaryloxyphosphazene

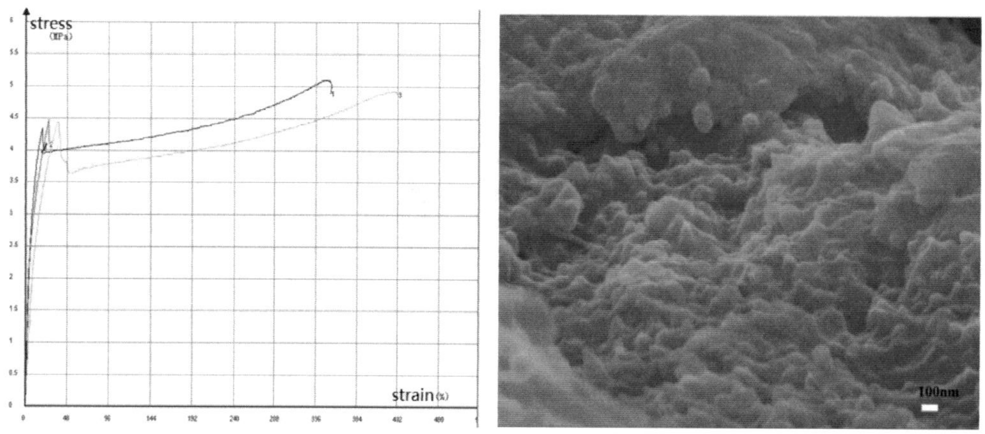

Figure 10. Stress-strain curves and cross-section SEM imagine of vulcanized samples

Table 4. Mechanical properties of vulcanized samples

Mechanical Properties	Tensile Strength	Elongation at break
values	5±0.3 MPa	270%~350%

4. Conclusion

In this report, a series of polyphosphazenes with different substitution rates of ethoxy were prepared, the measured values as well as theoretical calculated values of Tg were determined by DSC and Fox equation, respectively.

It found that a plot of the Tg vs. ethoxy contents yielded straight lines when polymer chain owned a higher substitution rate of phenoxy or ethoxy, which is in accordance with the deduced equation from Tailor-Gorden equation.

The simple method could be used to determine proportions of polymer structural units that match a certain target Tg value based on the linear fit relationships and guide the manufacture of the polyaryloxyphosphazene elastomer with better mechanical properties and low temperature resistance.

The obtained theoretical relationships can be likely to have broad applications in high performance polymers preparation, especially polyphosphazene chemistry. However, the major limitations in this work were insufficient extension of the relationships, which only confirmed the applications for polyphosphazene. The relevant works on applications for other polymers with structure units would be carried out in the future.

References

[1] H. R. Allcock, 2016, The expanding field of polyphosphazene high polymers, *Dalton. Trans.* **45**, 1856.

[2] H. R. Allcock, 2012, Polyphosphazene elastomers, gels, and other soft materials, *Soft. Matter.* **8** (29): 7521-7532.

[3] C. S. Reed, J. P. Taylor, K. S. Guigley, M. M. Coleman, H. R. Allcock, 2000, Polyurethane/ poly[bis(carboxylatophenoxy)phosphazene] blends and their potential as flame-retardant materials, *Polym. Eng. Sci.* **40** (2): 465-472.

[4] H. R. Allcock, R. Prange, 2001, Properties of poly(phosphazene-siloxane) block copolymers synthesized via telechelic polyphosphazenes and polysiloxane phosphoranimines, *Macromolecules.* **34** (20): 6858-6865.

[5] C. J. Orme, M. K. Harrup, J. D. Mccoy, D. H. Weinkauf, F. F. Stewart, 2002, Pervaporation of

water-dye, alcohol-dye, and water-alcohol mixtures using a polyphosphazene membrane, *J. Membrane. Sci.* **197** (1-2): 89-101.

[6] I. Manners, 1996, Polymers and the periodic table: recent developments in inorganic polymer science, *Angew. Chem. Int. Ed. Engl.* **27** (15): 1602-1621.

[7] H. R. Allcock, R. L. Kugel, K. J. 1966, Valan, Phosphonitrilic compounds. VI. High molecular weight poly(alkoxy-and aryloxyphosphazenes), *Inorg. Chem.* **5** (10): 1709-1715.

[8] R. E. Singler, G. L. Hagnauer, N. S. Schneider, B. R. Laliberte, R. E. Sacher, R. W. Matton, 1974, Synthesis and characterization of polyaryloxyphosphazenes, *J. Polym. Sci. Polym. Chem. Ed.* **12** (2): 433-444.

[9] J. L. Nichol, I. T. Hotham, H. R. Allcock, 2014, Ethoxyphosphazene polymers and their hydrolytic behavior, *Polym. Degrad. Stabil.* **109**: 92-96.

[10] D. P. Tate, 1974, Polyphosphazene elastomers, *J. Polym. Sci. Polym. Symposium.* **48** (1): 33-45.

[11] S. Zhang, S. Ali, Y. Kong, H. Ma, L. Zhang, Z. Wu, D. Wu, 2017, Evolution of chemical structure of polydichlorophosphazene in various solvents and ways to prolong its stability, *J. Mol. Liq.* **225**: 536-543.

[12] S. Zhang, S. Ali, H. Ma, L. Zhang, Z. Wu, D. Wu, T. S. Hu, 2016, Preparation of poly (bis(phenoxy)phosphazene) and ^{31}P-NMR analysis of its structural defects under various synthesis conditions, *J. Phys. Chem. B.* **120** (43): 11307-11316.

Fatigue property of cerium oxide modified vulcanised natural rubber under uniaxial loading

Wuwei Duan[1], Yanping Wang[1,3], Yazhou Liu[1], Zhe Zhang [2], Zheng Pan[1] and Yu Qiao[1]

[1]School of Chemical Engineering, Inner Mongolia University of Technology, Hohhot, China
[2]School of Chemical Engineering and Technology, Tianjin University, Tianjin, China

[3]Corresponding Author's E-mail: ypwang@imut.edu.cn

Abstract. Cerium oxide (CeO_2) modified natural rubber (NR) exhibits good mechanical properties, such as high abrasion resistance, high tear strength and good thermal aging and creep resistance. Rubber components are often subjected to long-term cyclic loading in service. Therefore, fatigue performance needs to be considered for their application. In the investigations, tensile properties and fatigue properties of CeO_2 modified vulcanised NR under uniaxial loading were investigated. The fatigue tests were carried out under constant cyclic strain range at room temperature. Results showed that the CeO_2 modified NR demonstrated higher elastic modulus and larger elongation than that of unmodified rubber, which were related to the enhanced interaction and bonding force of rubber chains by CeO_2 additive. Moreover, a remarkably enhanced fatigue life could be achieved in the CeO_2 modified NR. The internal friction and impact in the rubber matrix was relieved by CeO_2 additive, which can reduce the hysteresis loss during cyclic loading, resulting in improved fatigue resistance in CeO_2 modified vulcanised NR.

1. Introduction

Rare earth (RE) has been used extensively as reinforcing filler in the preparation of composites for its unique magnetic, luminescent, and electrochemical properties. RE can be applied to reduce weight, increase durability and thermal stability [1-4]. Therefore, RE compounds have been widely used in many devices, such as computer memory, cell phones, magnets, catalytic converters, fluorescent lighting and so on.

In recent years, an improvement of mechanical properties and thermal stability has been reported in the RE modified natural rubbers (NR). Guo et al. [5] studied the effects of three types of RE on the mechanical properties of the modified high-abrasion furnace-filled powdered natural rubbers (NR/HAF). Tensile strength and 300% modulus were increased by approximately 10% in the samarium oxide (Sm_2O_3) modified NR/HAF. And lanthanum oxide (La_2O_3) and praseodymium oxide (Pr_6O_{11}) could help to increase tear strength. Lin et al. [6] investigated effects of ten types of rare earth additives on mechanical properties of NR/HAF. The tear strength was improved obviously in all of RE modified NR/HAF. Moreover, RE was also beneficial for improving thermal stability of rubbers. La modified antioxidant could enhance thermo-oxidative aging resistance of NR composites. The expoxidized natural rubber (ENR) filled with 1 phr neodymium stearate (NdSt) exhibited improved thermal-oxidative stability compared to unmodified ENR [7-9].

Cerium oxide (CeO_2) is an important RE additive whose main application is for polishing, glass and catalytic converters. Li et al. [10, 11] studied systematically the influences of CeO_2 particles on mechanical properties of NR. The abrasion resistance could be enhanced by approximately 20% in the CeO_2 modified NR. It was valuable to note that the tear strength of NR filled with 1.226 m or 0.39 m CeO_2 particles was almost twice as much as that of unmodified NR. In addition, the enhanced thermal aging resistance was also achieved in the CeO_2 modified NR. Moreover, stress relaxation resistance and creep resistance were enhanced in the CeO_2 modified vulcanised NR, which were attributed to the enhanced interaction and bonding force of rubber chains caused by CeO_2 additive [12]. Therefore, it is worth emphasizing that CeO_2 particles could be used effectively as a reinforcing additive to enhance mechanical properties of NR, such as high abrasion resistance, high tear strength, good thermal aging resistance and creep resistance [13].

NR has been widely used to fabricate the automobile tires, vibration isolators, seals and gaskets in industry [14]. Since rubber components are frequently subjected to long-term cyclic mechanical load in service, therefore long-term durability becomes critical for their applications [15, 16]. Therefore, it is imperative to reveal the fatigue property of RE modified NR for their practical application. Carbon black and other additives were used as reinforcing fillers to improve fatigue resistance of rubbers [17-20]. The effects of additives on fatigue behavior were attributed to crack tip blunting, deviation and branching at the crack tip caused by inhomogeneity of the rubber-filler composites [21]. Liu et al. [18] demonstrated that hybrid carbon black and nanoclay could construct a dual phase filler network in NR, which could cause anisotropy in the matrix and introduce more energy dissipation mechanisms to the system, resulting in enhanced fatigue resistance. The relatively high fatigue resistance was also related to low hysteresis loss and energy input [19]. In addition, uniaxial ratcheting behavior of the CeO_2 modified vulcanised NR was studied under cyclic asymmetric stress-controlled experiments at room temperature [21].

So far, the influences of RE additives on fatigue properties of rubbers have not yet been investigated in detail. It would be of interest to understand the influence of RE additives on fatigue properties of rubbers. In the present work, the tensile properties and fatigue performance of the CeO_2 modified vulcanised NR are presented. Moreover, the fatigue fracture mechanism is also discussed.

2. Experiment

CeO_2 particles were provided by Jilin Chemical Industrial Limited Company of China. As shown in Fig. 1, the powder shape is irregular, and particle size ranges from approximately 0.5 μm to 5 μm. The formulations of the CeO_2 modified vulcanised NR are shown in Table 1. The symbols NR and NR/CeO_2 present unmodified rubber and the rubber filled with 2 phr CeO_2, respectively.

Fig.1. The morphology of CeO_2 particles.

Table 1. The formulations of the CeO_2 modified vulcanised natural rubber

Ingredient	Composition, phr
NR	100.0
Carbon Black	68.0
Zinc Oxide	5.0
Softeners	2.0
Activator	2.0
Accelerator	0.5
Antioxidant	4.0
Sulfur	1.8
Cerium Oxide	2

Tensile tests were carried out on a universal testing machine under displacement control with cross-head speed of 5mm/min. Maximum strain criterion was used to predict fatigue life of the rubbers [17, 22-24]. Cyclic response and fatigue tests were carried out using a minitype-tension-torsion fatigue machine at room temperature under displacement control. A triangle waveform loading with a constant strain rate of 2/s under pulse cyclic tension was used in fatigue tests. Specimens were subjected to five levels of maximum strain at room temperature: 100%, 125%, 150%, 175% and 200%. The point of inflection of load drop was termed as fatigue life. At least three specimens were tested for each condition to verify the reproducibility of the results. The specimens corresponded to a hollow-dumbbell-shaped had a gauge length, outer diameter, and wall thickness of 22 mm, 8 mm, and 1 mm, respectively. The detailed dimensions of the specimens for fatigue tests are shown in Fig.2. In order to keep the hollow shape, grip section of the specimen was filled with steel stick [21, 24, 25].

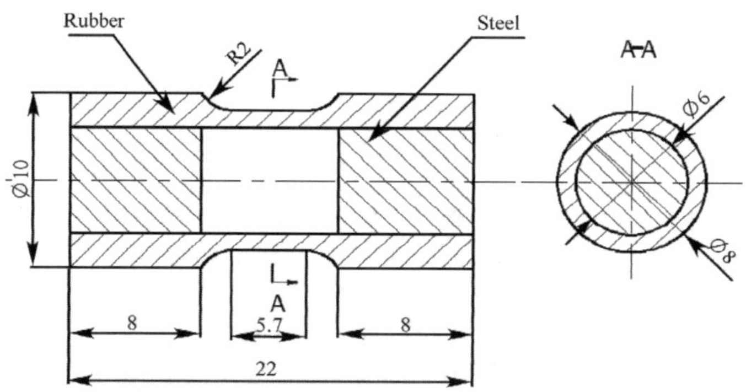

Fig. 2. The dimensions of the specimens for fatigue tests.

The morphology of CeO_2 particles and fracture surface were characterized by scanning electron microscopy (SEM, Hitachi SU4800) coupled with energy dispersive X-ray spectroscopy (EDS). The fracture surface was previously coated with a thin gold layer by vapor deposition before observation. The crosslink density of the rubbers was measured by a nuclear magnetic resonance (NMR) crosslink density instrument (MR-CDS3500).

3. Results and discussion

3.1. Microstructure and mechanical properties of the CeO_2 modified vulcanised NR

Figure 3 shows the distribution of CeO_2 particles in the modified vulcanised NR. Non-agglomerated particle was embedded in the rubber matrix. The mechanical properties of the CeO_2 modified vulcanised NR are summarized in Table 2. It is found that the differences in tensile strength, stress at 200% elongation and hardness between NR and NR/CeO_2 are insignificant. However, the elastic modulus and elongation at break of NR increases in the NR/CeO_2 compared to unmodified sample. The elastic modulus of unmodified sample is 2.1 MPa. By contrast, this value of NR/CeO_2 increases by 24%. In addition, the elongation at break of NR is approximately 466%, while that of NR/CeO_2 is approximately 526%. Therefore, the elastic modulus and elongation are enhanced in the NR/CeO_2. Similar tendency in mechanical properties of the CeO_2 modified rubbers was also reported by Lin [6]and Li[26]. Moreover, the crosslink density of NR/CeO_2 is approximately 1.385×10^{-4} mol/cm^{-3}, which is higher than that of unmodified sample (1.242×10^{-4} mol/cm^{-3}).

Table 2. Mechanical properties of the CeO_2 modified vulcanised natural rubber.

Samples	Crosslink density (mol/cm^{-3})	Elastic modulus (MPa)	Tensile Strength (MPa)	Elongation at break (%)	Stress at 200% elongation (MPa)	Degree of hardness (A type)
NR	1.242×10^{-4}	2.1	20.9	471	5.2	66
NR/CeO_2	1.385×10^{-4}	2.5	21.7	526	5.2	64

RE has a large quantity of empty orbits due to the electron shell structure, which can bind or capture the free radicals generated from the autocatalytic, free radical chain reactions the steric hindrance effect of the formed complex structure. It is known that the structure of CeO_2 has a lot of vacant 'f' orbit, and complex compound tends to form in the CeO_2 modified rubber matrix [26]. CeO_2 additive could enhance the interaction and bonding force of rubber chains, which causes the increased crosslink density, elastic modulus and elongation.

3.2. Fatigue properties of the CeO_2 vulcanised NR

Figure 4 shows the stress-strain responses of the CeO_2 modified and unmodified rubbers at a maximum strain of 200% during the first 10 cycles. It can be observed that both modified and unmodified samples show a large degree of softening resulting in strong nonlinear unloading curves, which is termed as Mullins effect [27]. As shown in Fig.4, both CeO_2 modified and unmodified rubbers show similar stable stress-strain response.

Figure 5 shows the variation in peak engineering stress values of the CeO_2 modified and unmodified vulcanised NR. Stress softening can be observed in both NR and NR/CeO_2, which was explained by change of rubber's stiffness during cyclic deformation [22]. However, it is worth emphasizing that the improvement of fatigue life can be achieved in the modified rubber filled with CeO_2 additive. In order to illustrate the influences of CeO_2 additive on the improvement of fatigue life more clearly, the maximum strain ranging from 100% to 200% is plotted as a function of fatigue life in Fig. 6. It can be noted that the fatigue life of both CeO_2 modified and unmodified samples significantly reduces as strain range increases from 100% to 200%. However, fatigue life reduces slowly when the strain range is larger than 150%. As indicated in Fig.6, the fatigue life of NR/CeO_2 increases by approximately 50% compared to unmodified sample under the same cyclic strain range. Therefore, the fatigue life of vulcanised NR could be remarkably improved by filling CeO_2 additive.

The strain energy density, which is calculated from the area within the stable hysteresis loop, has been used extensively as a damage parameter to characterize fatigue failure during cyclic deformation [28]. In the present study, the strain energy density is calculated from the 200th cycle, wherein stress-strain response tends to be stable. Figure 7 presents the variation of strain energy density of NR and NR/CeO_2. It can be observed that the strain energy density decreases rapidly during the first 2000 cycles. Afterwards, it tends to be stable in each cycle. It is interesting to note that the strain energy density of the modified rubber filled with CeO_2 additive is remarkably lower than that of unmodified rubber. It is known that hysteresis is caused by internal friction in rubber matrix. As indicated in Fig.7, the strain energy density decreased in the NR/CeO_2 compared to unmodified rubber, which indicates that the internal friction in rubber matrix can be relieved by CeO_2 additive, resulting in reduction of hysteresis loss.

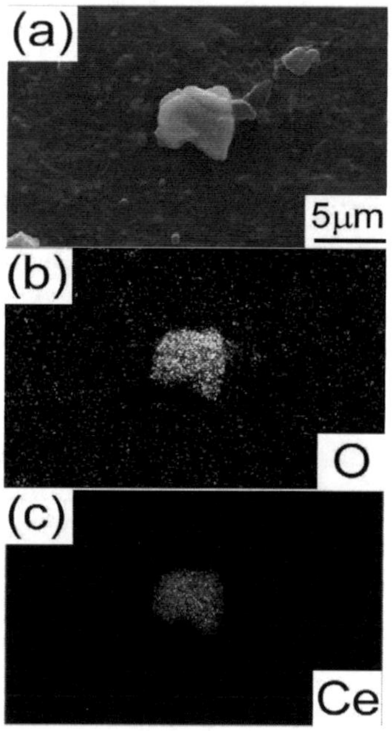

Fig. 3. The morphology of CeO_2 particle in the modified vulcanised natural rubber: (a) Morphology of CeO2 particles; (b) Oxygen map; (c) Cerium map.

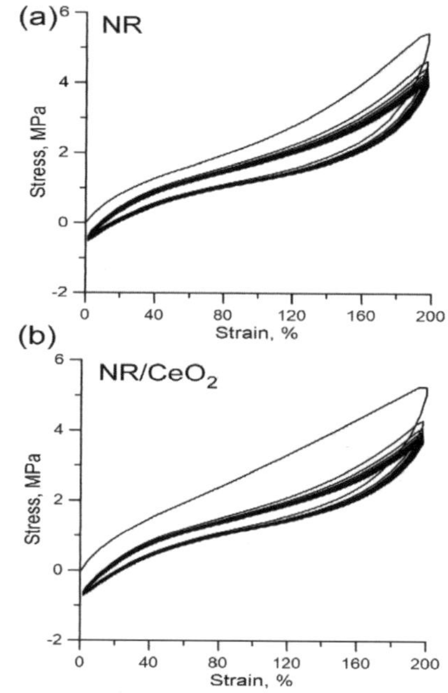

Fig. 4. Stress-strain responses of the rubbers at maximum strain of 200% under cyclic loading (first 10 cycles): (a) NR; (b) NR/ CeO_2.

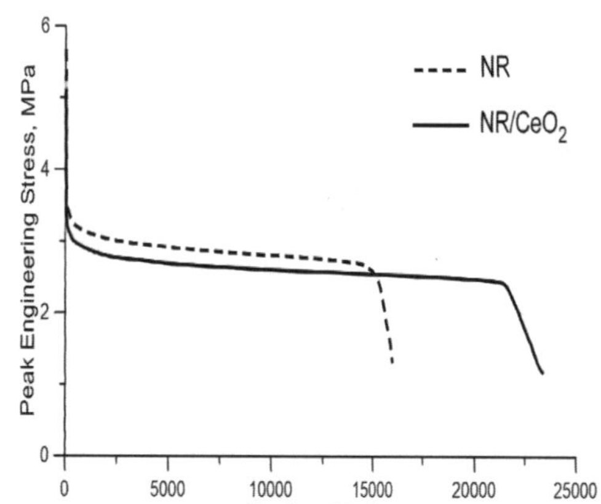

Fig. 5. Variation of peak stress of the CeO_2 modified and unmodified rubbers at maximum strain of 200% under cyclic loading.

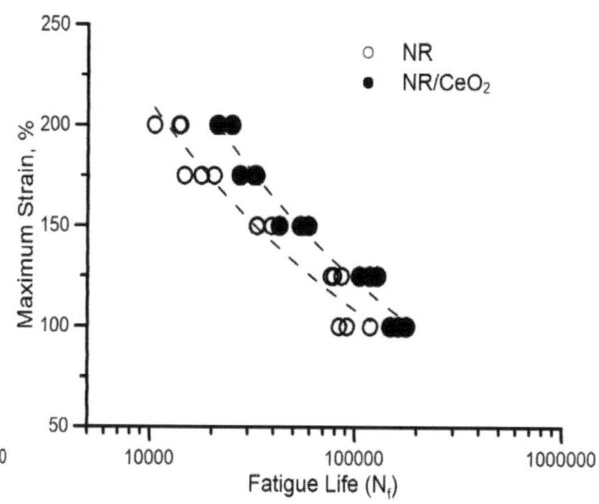

Fig. 6. Fatigue life of the CeO_2 modified and unmodified rubbers under various strain ranges.

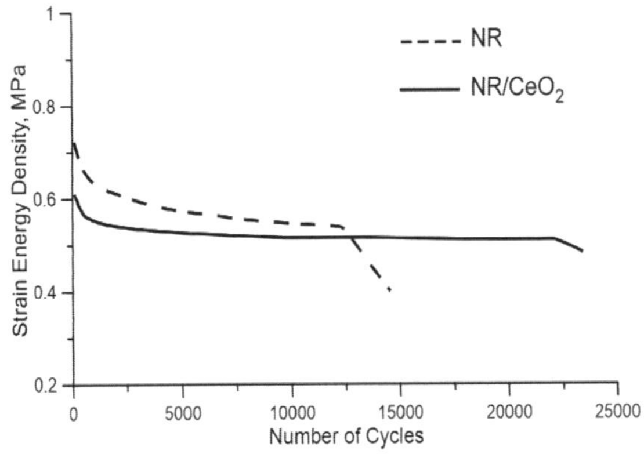

Fig. 7. Variation of strain energy densities of the CeO_2 modified and unmodified rubbers at maximum strain of 200%.

3.3. Fatigue fracture mechanism of the CeO_2 modified vulcanised NR

The typical morphology of fatigue fracture surface of NR and NR/CeO_2 is shown in Fig.8. The fracture surfaces show the fatigue characteristic features, wherein the crack initiation, crack propagation zone and the rapid fracture zone can be easily identified. As indicated in Fig.8a and 8b, macroscopic fatigue cracks in both NR and NR/CeO_2 initiate from specimen surface, and several visible particles can be observed in the crack initiation region. It is well-known that lots of oxide particles have been used as softener, activator, accelerator and antioxidant in rubber products. These particles are often harder than rubber matrix, which tends to induce stress concentration near these particles during fatigue tests that are identified as defect in rubber matrix resulting in crack initiation. Therefore, fatigue cracks initiate near these additives in both CeO_2 modified and unmodified samples.

Fatigue fracture surface typically consists of two distinct phases, the rough region and smooth region. The rough region corresponds to fatigue crack propagation zone, wherein wrenching oriented in the direction of the defect can be observed. Since emerging cracks propagate in specimen depth, the wrenchings form an arc shape. This fracture behavior was explained by inhomogeneous local stress distribution and strain crystallization [24, 29]. The wrenchings shape and size are similar in both NR and NR/CeO_2. By contrast, as shown in Fig.8c and 8d, the smooth region indicates rapidly crack propagation, corresponding to final stage of fatigue fracture [29]. It is clear that rapid fracture zone of NR/CeO_2 is smaller than that of NR.

Rubber balling was reported in the fatigue induced carbon black filled natural rubber and nitrile rubber [22, 30]. The formation of rubber balling was recognized as the friction and impact of the simultaneous effect of continuous cyclic loading on the interaction between carbon black and the molecular chains. Plenty of rubber balls are distributed randomly in the crack propagation zone as shown in Fig.8. However, it is interesting to note that the size and amount of rubber balls are different in NR and NR/CeO_2. The size of rubber balls in NR ranges from 10 m to 70 m, which is larger than that in NR/CeO_2. Moreover, compared to NR, NR/CeO_2 has fewer amount of rubber balls. Therefore, NR/CeO_2 demonstrates weaker internal friction and impact between carbon black and the molecular chains compared to unmodified sample, resulting in reduction of hysteresis loss during fatigue. This phenomenon is consistent with the decline of strain energy density obtained in NR/CeO_2 during fatigue.

As illustrated in previous studies [6, 31], the CeO_2 modified rubbers exhibited combination of high tear strength, high abrasion resistance, good thermal aging resistance and creep resistance. Due to internal friction and impact in the rubber matrix restrained by CeO_2 additive during cyclic loading, the enhanced resistance to fatigue achieved in NR/CeO_2 is attributed to its low hysteresis loss. Based on our investigations, CeO_2 is an effect additive to enhance fatigue resistance of NR. The improved long-term durability in CeO_2 modified rubber is attractive for its practical applications.

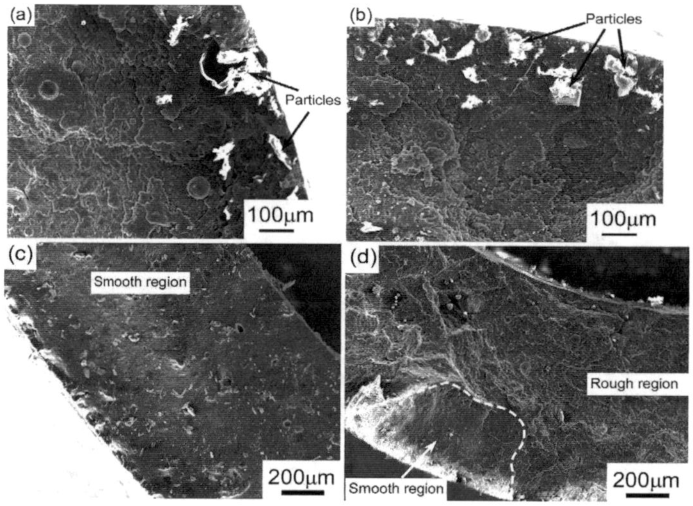

Fig. 8. Morphology of fatigue fracture surface: (a) Crack initiation area in NR; (b) Crack initiation area in NR/CeO_2; (c) Crack propagation area in NR; (d) Crack propagation area in NR/CeO_2

4. Conclusions

In the present study, the cyclic deformation and fatigue properties of the CeO_2 modified vulcanised NR were studied under uniaxial mechanical loading. Based on these analyses, the following conclusions can be outlined:

(1) The crosslink density of the modified vulcanised NR could be increased by CeO_2 additive. The CeO_2 modified vulcanised NR show higher elastic modulus and larger elongation compared to unmodified rubber.

(2) A remarkably enhanced fatigue life could be achieved in the rubber filled with CeO_2 additive. The internal friction and impact in the rubber matrix can be relieved by CeO_2 additive, which reduces the hysteresis loss during fatigue test, resulting in improvement of fatigue resistance in CeO_2 modified vulcanised NR.

(3) CeO_2 is an effect additive to enhance the fatigue resistance of vulcanised NR. Due to the combination of high tear strength, high abrasion resistance, good thermal aging resistance and good fatigue resistance, CeO_2 modified vulcanised NR demonstrated good application potential in industry.

Acknowledgment

The authors are grateful for the financial support from the National Natural Science Foundation of China (Grant No. 11562015, 51605325) and Science and Technology on Reactor System Design Technology Laboratory.

References

[1] S. Yang, L. Liu, Z. Jia, D. Jia, Y. Luo, Structure and mechanical properties of rare-earth complex La-GDTC modified silica/SBR composites, *Polymer*, **52** (2011) 2701-2710.

[2] W. Zheng, L. Liu, X. Zhao, J. He, A. Wang, T.W. Chan, S. Wu, Data for effects of lanthanum complex on the thermo-oxidative aging of natural rubber, *Polymer Degradation & Stability*, (2015).

[3] C. Sun, H. Li, L. Chen, ChemInform Abstract: Nanostructured Ceria-Based Materials: Synthesis, Properties, and Applications, **43** (2012).

[4] X. He, D. Zhang, H. Li, J. Fang, L. Shi, Shape and size effects of ceria nanoparticles on the impact strength of ceria/epoxy resin composites, *Particuology*, (2011) 80-85.

[5] Y.B. Guo, L.S. Wang, A.Q. Zhang, Mechanical properties and crosslink density of rare earth-modified high-abrasion furnace-filled powdered natural rubber, *Journal of Applied*

Polymer Science, **102** (2006) 1755-1762.
[6] Y.L. Lin, A.Q. Zhang, L.S. Wang, Rare earth compounds modified carbon black filled powdered natural rubber: Preparation, morphology and properties, *Journal of Applied Polymer Science*, **108** (2008) 1393-1401.
[7] L. Fang, Y.H. Song, X.N. Zhu, Q. Zheng, Influence of lanthanum stearate as a co-stabilizer on stabilization efficiency of calcium/zinc stabilizers to polyvinyl chloride, *Polymer Degradation and Stability*, **94** (2009) 845-850.
[8] Y.Y. Luo, C.J. Yang, B.Q. Chen, K. Xu, J.P. Zhong, Z. Peng, Z. Lu, Q.H. Wang, Thermal degradation of epoxidized natural rubber in presence of neodymium stearate, *Journal of Rare Earths*, **31** (2013) 526-530.
[9] W. Zheng, L. Liu, X.Y. Zhao, J.W. He, A. Wang, T.W. Chan, S.Z. Wu, Effects of lanthanum complex on the thermo-oxidative aging of natural rubber, *Polymer Degradation and Stability*, **120** (2015) 377-383.
[10] L.H. Li M, Study on Cerium Oxide Modified Natural Rubber, *Journal of Rare Earths*, (2007) 138-141.
[11] L. Mei, S. Zhenxue, L. Zhaogang, H. Yanhong, W. Mitang, L. Hangquan, Effect of Surface Modification on Behaviors of Cerium Oxide Nanopowders, *Journal of Rare Earth*, (2007).
[12] Y. Wang, Y. Liu, Z. Zhang, C. Wang, S. Shi, X. Chen, Mechanical properties of cerium oxide-modified vulcanised natural rubber at elevated temperature, *Plastics Rubber & Composites*, (2017) 1-8.
[13] W. Duan, Y. Wang, C. Ji, Z. Zhang, X. Chen, J. Liu, Z. Pan, Y. Qiao, Microstructure of uniaxially fatigued thermally aged vulcanised natural rubber filled with cerium oxide, *Journal of Rubber Research*, (2020).
[14] W.V. Mars, A. Fatemi, A literature survey on fatigue analysis approaches for rubber, *International Journal of Fatigue*, **24** (2002) 949-961.
[15] A. Fatemi, N. Shamsaei, Multiaxial fatigue: An overview and some approximation models for life estimation, *International Journal of Fatigue*, **33** (2011) 948-958.
[16] Y.C. Lin, X.M. Chen, J. Zhang, Uniaxial ratchetting behavior of anisotropic conductive adhesive film under cyclic tension, *Polymer Testing*, **30** (2011) 8-15.
[17] J.H. Kim, H.Y. Jeong, A study on the material properties and fatigue life of natural rubber with different carbon blacks, *International Journal of Fatigue*, **27** (2005) 263-272.
[18] Y.B. Liu, L. Li, Q. Wang, X. Zhang, Fracture properties of natural rubber filled with hybrid carbon black/nanoclay, *J Polym Res*, **18** (2011) 859-867.
[19] Y.J. Nie, L.L. Qu, G.S. Huang, B.Y. Wang, G.S. Weng, J.R. Wu, Improved resistance to crack growth of natural rubber by the inclusion of nanoclay, *Polymers for Advanced Technologies*, **23** (2012) 85-91.
[20] B. Dong, C. Liu, L.Q. Zhang, Y.P. Wu, Preparation, fracture, and fatigue of exfoliated graphene oxide/natural rubber composites, *Rsc Advances*, **5** (2015) 17140-17148.
[21] S.K. Li, H. Wang, Y.P. Wang, C. Wang, H.G. Niu, J.L. Yang, Uniaxial ratchetting behaviour of cerium oxide filled vulcanized natural rubber, *Polymer Testing*, **32** (2013) 468-474.
[22] M.S. Loo, J.B. Le Cam, A. Andriyana, E. Robin, A.M. Afifi, Fatigue of swollen elastomers, *International Journal of Fatigue*, **74** (2015) 132-141.
[23] M. Flamm, J. Spreckels, T. Steinweger, U. Weltin, Effects of very high loads on fatigue life of NR elastomer materials, *International Journal of Fatigue*, **33** (2011) 1189-1198.
[24] Y.P. Wang, X. Chen, W.W. Yu, Microscopic mechanism of multiaxial fatigue of vulcanised natural rubber, *Plast Rubber Compos*, **40** (2011) 491-496.
[25] Y. Wang, W. Yu, X. Chen, L. Yan, Fatigue life prediction of vulcanized natural rubber under proportional and non-proportional loading, *Fatigue & Fracture of Engineering Materials & Structures*, **31** (2008) 38-48.
[26] L. Mei, S. Zhenxue, L. Zhaogang, H. Yanhong, W. Mitang, L. Hangquan, Effect of Surface Modification on Behaviors of Cerium Oxide Nanopowders, *Journal of Rare Earth*, **25** (2007).

[27] J. Diani, B. Fayolle, P. Gilormini, A review on the Mullins effect, *Eur Polym J*, **45** (2009) 601-612.
[28] B. Dong, C. Liu, Y.P. Wu, Fracture and fatigue of silica/carbon black/natural rubber composites, *Polymer Testing*, **38** (2014) 40-45.
[29] J.B. Le Cam, B. Huneau, E. Verron, Fatigue damage in carbon black filled natural rubber under uni- and multiaxial loading conditions, *International Journal of Fatigue*, **52** (2013) 82-94.
[30] A.K. Bhowmick, S. Basu, S.K. De, Scanning electron microscopy studies of abraded rubber surfaces, *Journal of Materials ence*, **16** (1981) 1654-1660.
[31] M. Li, Z.X. Shi, Z.G. Liu, Y.H. Hu, M.T. Wang, H.Q. Li, Effect of surface modification on behaviors of cerium oxide nanopowders, *Journal of Rare Earths*, **25** (2007) 368-372.

IOP Conf. Series: Materials Science and Engineering 1167 (2021) 012005 doi:10.1088/1757-899X/1167/1/012005

Characteristics and mechanism of light-induced resistance effect in ZnO thin film/SiO_2/Si heterojunction

X Y Zhou[1,2], B S Yin[1]

[1]College of Science, China University of Petroleum, Qingdao, 266580, Shandong, P. R. China

[2]Corresponding Author's E-mail: zhouxiaoyan@upc.edu.cn

Abstract. The polycrystalline ZnO thin films have been deposited on p-Si(100) substrate with a native SiO_2 layer by radio frequency magnetron sputtering. The photoinduced position sensitivity of the ZnO thin film/SiO_2/Si heterojunction has been investigated. The resistance relative change ratio of ZnO/SiO_2/Si was about 1041% at 1.0 bias voltage when the different positions of the heterojunction surface were illuminated by a He-Ne laser with 5 mW power and 50 μm diameter spot. A gradient distribution of light-generated electrons in the sample varied with the location of the light spot. The reason was that the distribution of carrier density is different between the illuminated and unilluminated location.

1. Introduction

The effective control of the resistance of the device has been of great concern as data storage technology, memory switch and molecular optoelectronic switches [1-5]. The common methods are by changing temperature or applying an external field, such as superconductivity [6-10] and giant magneto resistance (GMR) [11-13]. Recently, Wang et al. reported a new bipolar-resistance effect based on metal-oxide-semiconductor (MOS) structure of Ti/SiO_2/Si, Co/SiO_2/Si and Cu/SiO_2/Si [11]. Chen et al. reported photoinduced modulation in resistance switching of carbon nanotube/$BiFeO_3$/Pt heterostructure [15]. They pointed out that the dual asymmetric junctions in the heterojunction could be addressed as a critical factor. Shablaev et al. reported effect of light-induced drop in electrical resistance of bulk $SrTiO_3$ crystals [16]. Carbon materials have been reported to be effective in manipulating resistors such as single-walled carbon nanotubes, double-walled carbon nanotubes and carbon film [17-21]. However, there are few studies on the photoinduced position sensitivity of the resistance of ZnO thin film/SiO_2/Si heterojunction. In this work, we report the discovery of a novely light-induced resistance effect in the ZnO thin film/SiO_2/Si structure.

2. Experiment

The polycrystal ZnO thin films were fabricated on the surface of p-type monocrystalline Si using a radio-frequency (RF) magnetron sputtering at room temperature, in which the resistivity of Si is 12-18 Ω·cm and the diffraction crystal plane is (100). The surface of p-Si was covered by a very thin natural oxide layer of SiO_2. It was cleaned in acetone, ethanol and deionized water for 5-10 minutes in turn using an ultrasonic method. After that, the Si substrate was blown dry by nitrogen and transferred to the sputtering chamber. A high pure ZnO ceramic target (diameter of 6 cm) was used. The base pressure of the sputtering chamber is less than 1.6×10^{-4} Pa before the deposition of the film. Then the surface contamination of the Si substrate was removed by pre-sputtering process. During the growth of

ZnO film, RF power was fixed at 100 W and the deposition pressure was kept at 1 Pa. The sputtering gas was a mixture of argon and oxygen, and the volume ratio of argon to oxygen is 3:1. Two translucent indium (In) electrodes with a distance of 7 mm were deposited on the surface of ZnO films by direct current magnetron sputtering.

The crystalline structure features of ZnO thin film was characterized by X-ray diffractometer (XRD), using a DX-2700 X diffractometer with wavelength radiation of 1.5409 Å (Cu K_α) and diffraction angle in the range of 20-60°. The surface morphology was analyzed by S4800 field-emission scanning electron microscope (SEM) with energy dispersive spectrometer (EDS) operated at 3.5 kV. The thickness of ZnO film was about 81 nm measured by spectroscopic ellipsometer. The direct resistance and current-voltage characteristics of ZnO/SiO_2/Si structure were measured by a Keithley 2400 sourcemeter using two-electrode method. The surface of samples was illuminated by a He-Ne laser with power of 5 mW and spot diameter of 50 μm.

3. Results and discussion

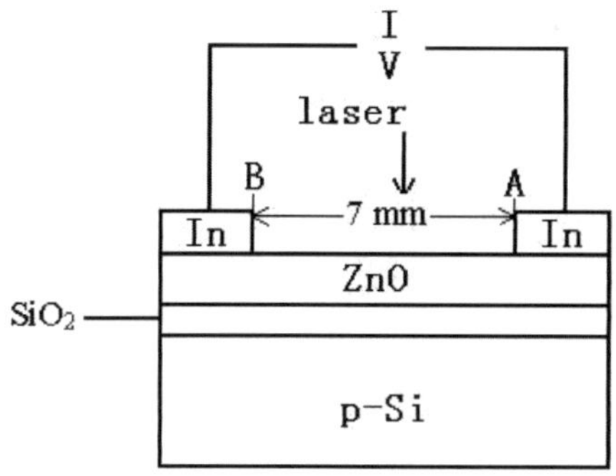

Figure 1. Schematic diagram of test structure of the ZnO thin film/SiO_2/Si heterojunction under He-Ne laser

As shown in Fig. 1, the He-Ne laser light was vertically on the surface of ZnO thin film/SiO_2/Si heterojunction. Two semitransparent metal In electrodes were deposited on the surface of ZnO thin film by direct current magnetron power. The thickness of In electrode was about 40 nm determined by spectroscopic ellipsometer.

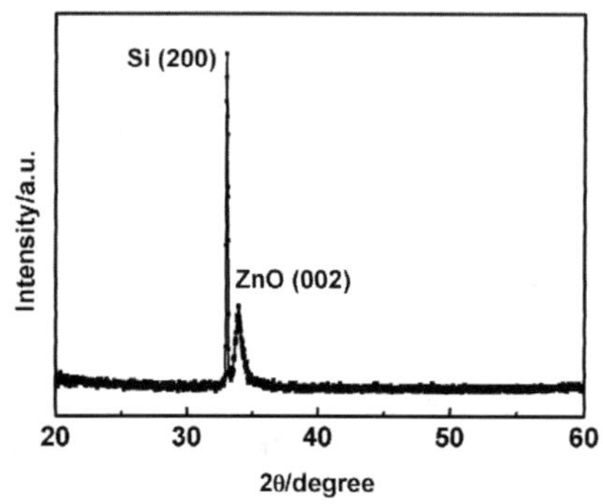

Figure 2. XRD spectra of the ZnO thin film/SiO_2/Si sample

The XRD patterns of the ZnO thin film/SiO_2/Si heterojunction are show in Fig. 2. The diffraction peak of ZnO (002) with strong Si (200) peak is clear distinguishable and indexed to hexagonal wurtzite ZnO (JCPDS card No. 79-2205, a =3.249 Å and c=5.205 Å). No other peaks from Zn and ZnO_2 impurities are observed. It indicates the highly c-axis-oriented growth and a high degree of crystallinity of ZnO thin film.

The surface morphology of ZnO thin film deposited on p-Si substrate is presented in Fig. 3(a). It shows that the ZnO thin film is continuous coatings with good adherence to the Si substrate, which consists of tightly packed round or oval particles. The average grain size is about 30 nm. EDX measurement was performed on the surface of $ZnO/SiO_2/Si$ heterojunction. The obtained results is shown in Fig. 3(b). The as-deposited $ZnO/SiO_2/Si$ heterojunction contains three elements, silicon, zinc and oxygen, and no other impurities. The silicon content is more than 80% because ZnO the film is very thin and the substrate is Si.

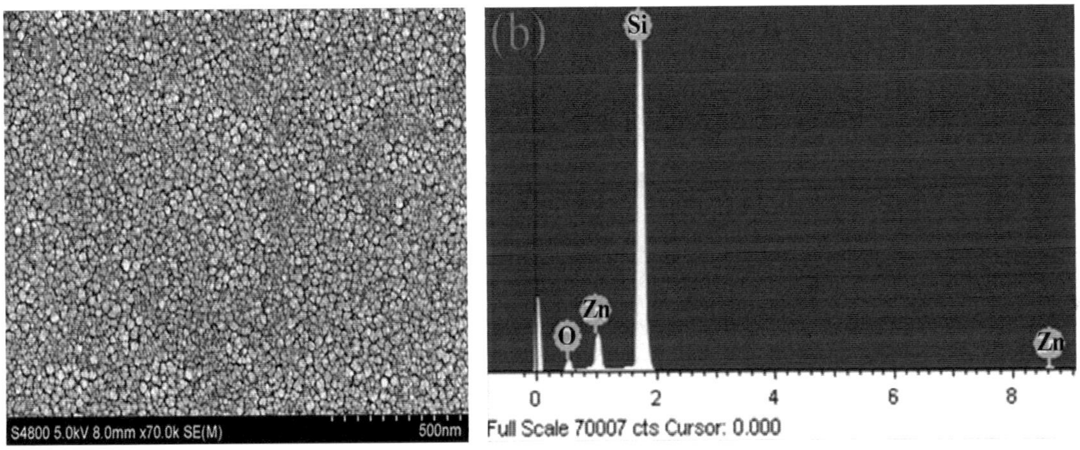

Figure 3. (a) SEM photograph of the ZnO thin film/SiO_2/Si sample, (b) EDX analysis of ZnO thin film deposited on p-Si substrate.

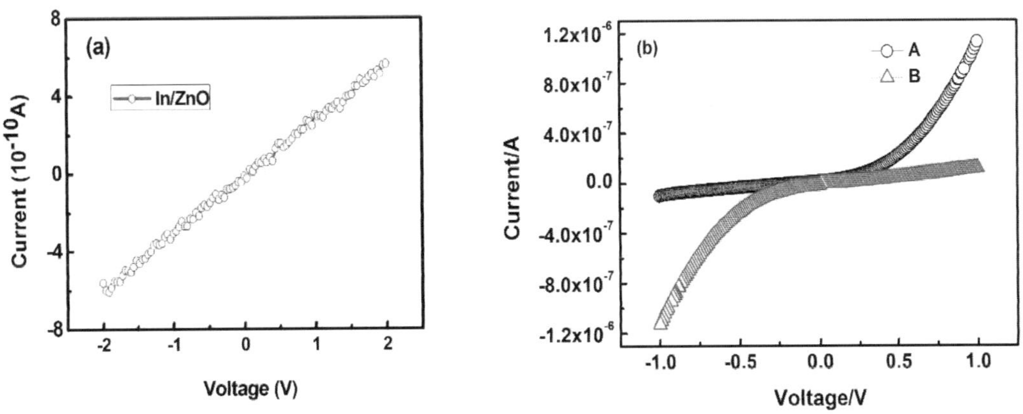

Figure 4. (a) *I-V* characteristics of the ZnO thin film, (b) *I-V* characteristics of the ZnO thin film/SiO_2/Si structure in laser position A and B

In order to prove that the contact between the In electrode and ZnO film is ohmic contact, the film is prepared on the substrate glass according to the original experimental conditions and parameters. As shown in Fig. 4(b), the *I–V* characteristic curve is basically straight which is found in other studies [22-23], confirming the contact ohmic contact between the In electrode and ZnO film. The photoinduced resistance sensitivity is defined by the following equation.

$$\Delta = \frac{R_{max} - R_{min}}{R_{min}} \times 100\% \quad (1)$$

In this location, the measurement bias voltage at position A is positive, and the voltage at position B is negative (A_+, B_-). Fig. 4(b) shows the *I-V* characteristics of the ZnO thin film/SiO_2/Si structure in laser position A and B. Interestingly, the curves of illuminating A and B are just opposite. When illuminating A, the resistance of heterojunction is reduced to the lowest value ($R_{min}=8.85\times10^5$ Ω) at forward voltage 1 V. The resistance of heterojunction is increased to the highest value ($R_{max}=1.01\times10^7$ Ω) when the laser illuminates at position B. The photoinduced resistance sensitivity of the heterojunction exhibits Δ= 1041%.

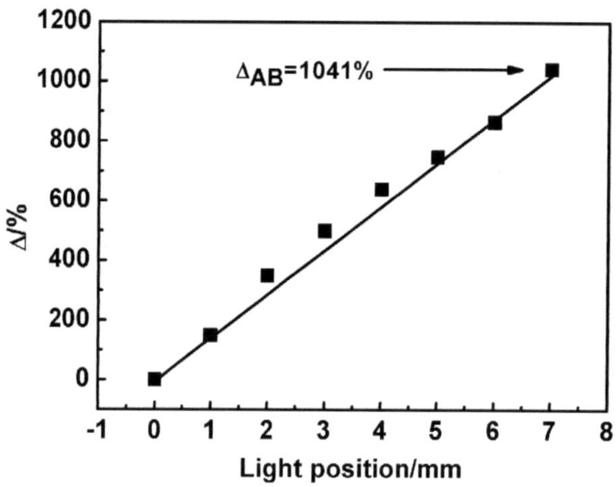

Figure 5. Resistance sensitivity change as a function of laser position on the ZnO thin film/SiO_2/Si heterojunction at 1.0 V bias voltage

Fig. 5 shows the relation of resistance change sensitivity and laser spot position on the surface of sample at 1.0 V bias voltage. The change ratio in resistance (Δ) increases with increasing the distance between A and B. The heterojunction exhibits an approximately linear relationship between the Δ and the laser spot displacement, which indicates that the ZnO thin film/SiO_2/Si heterojunction has potential application as precise photoelectric sensor.

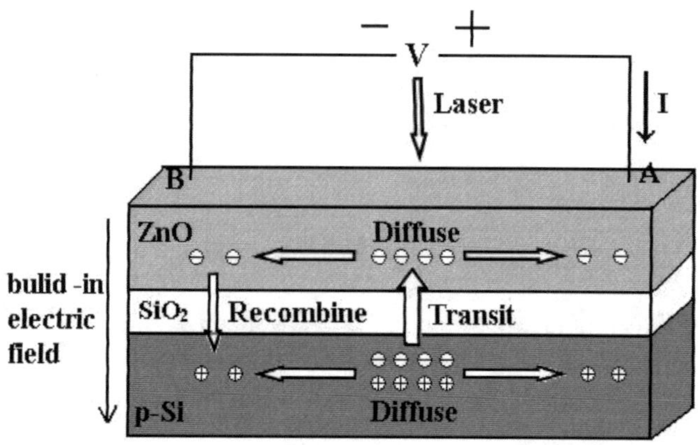

Figure 6. Schematic diagram of carrier diffusing in the ZnO thin film/SiO_2/Si heterojunction under laser illumination

Fig. 6 shows the schematic diagram of carriers diffusing in the ZnO thin film/SiO_2/Si heterojunction under laser irradiation. The electron mobility and carrier density of ZnO thin film which were measured by Hall effect measurement system are 4.5×10^3 cm^2/Vs and 8.1×10^{14} cm^{-3} at 300 K, respectively. When the He-Ne laser with 632.8 nm illuminates surface of the sample, light passes through the ZnO thin film to p-Si substrate and electron-hole are generated inside p-Si substrate. These photogenerated electrons will to be tunneled through SiO_2 layer into ZnO thin film by the build-in electric field. Some of the photogenerated electrons recombine with the holes and the rest are scattered in the ZnO film. There is a gradient distribution caused by excess electrons between the illuminated and unilluminated region. The direction of the external electric field is from A to B. Under the external electric field, it is easier for the excess electrons to move towards A, and more difficult to move towards B. So the heterojunction presents a low resistance at the right region, while presents a high resistance at the left region.

4. Conclusion

In summary, ZnO thin film/SiO_2/Si structure was prepared by radio frequency magnetron sputtering. Its resistance is very sensitive to the position of the laser. When the He-Ne laser with wavelength of 632.8 nm illuminated on the surface of the sample, the sample exhibited a large resistance change rate of approximately 1041%. The light-induced position of resistance sensitivity of the ZnO/SiO_2/Si is mainly attributed to the different carrier density distributions of different irradiation locations. The light-induced resistance effect is substantial and remarkable compare the other resistance effects. It shows that the ZnO thin film/SiO_2/Si heterojunction is an attractive candidate for photoelectric device.

Acknowledgments

This work was supported partly by the National Natural Science Foundation of China under Grant 51777215, partly by the Key Research and Development Plan Project of Shandong Province under Grant 2018GSF117002.

References

[1] Hopkins D S, Pekker D, Goldbart P M, et al, 2005 *Science* vol **308** p 1762
[2] Bezryadin A, Lau C N, Tinkham M, 2000 *Nature* vol **404** p 971
[3] Chang S H, Lee J S, Chae S C, et al, 2009 *Phys. Rev. Lett.* vol **102** p 026801
[4] Li J, Speyer G, Sankey O F, 2004 *Phys. Rev. Lett.* vol **93** p 248302
[5] Sirringhaus H, Tessler N, Friend R H, 1998 *Science* vol **280** p 1741
[6] Bardeen J, Cooper L N, Schrieffer J R, 1957 *Phys. Rev.* vol **108** p 1175
[7] Wu M K, Ashburn J R, Torng C J, et al, 1987 *Phys. Rev. Lett.* vol **58** p 908
[8] Nagamatsu J, Nakagawa N, Muranaka T, et al, 2001 *Nature* vol **410** p 63
[9] Onufrieva F, Pfeuty P, 2009 *Phys. Rev. Lett.* vol **102** p 207003
[10] Taylor B J, Maple M B, *Phys. Rev. Lett.* vol **102** p 137003
[11] Xu R, Husmann A, Rosenbaum T F, et al, *1997 Nature* vol **390** p 57
[12] Pandey A, Mazumdar C, and Ranganathan R, 2009 *Appl. Phys. Lett.* vol **94** p 172509
[13] Jin S, McCormack M, Tiefel T H, *1994 J. Appl. Phys.* vol **76** p 6929
[14] Yu C Q and Wang H, 2010 *Adv. Mater.* vol **22** p 966
[15] Chen Y, Zang Y Y, Xie D, 2012 *Integr. Ferroelectr. Int. J.* vol **134** p 1
[16] Shablaev S I and Grachev A I, 2014 *Phys. Solid State* vol **56** p 4
[17] Cai J Z, Lu L, Kong W J, et al, 2006 *Phys. Rev. Lett.* vol **97** p 026402
[18] Oshima Y, Takenobu T, Yanagi K, 2010 *Phys. Rev. Lett.* vol **104** p 016803
[19] Fedorov g, Lassagne B, Sagnes M, et al, 2005 *Phys. Rev. Lett.* vol **94** p 066801
[20] Ma M, Xie Q Z, Li L P, et al, 2011 *J. Appli. Phys.* vol **110** p 054503
[21] Wang B S, Tong P, Sun Y P, 2009 *Appl. Phys. Lett.* vol **95** p 222509
[22] Wang P, Chen N F, Yin Z G, 2006 *Appl. Phys. Lett.* vol **88** p 152102
[23] Gao X L, Xue Q Z, Hao L Z, et al, 2007 *Appl. Phys. Lett.* vol **91** p122110

Noval Ag·AgBr@TiO_2 porous power: fabrication and photocatalyst

Liangfa Gong[1,3], Zhe Wu[2], Yi Hao[2], Zhiyang Guo[2] and Cuiqing Li[1]

[1]College of Chemical Engineering, Beijing Institute of Petro-Chemical Technology, Beijing, China
[2]Applied Chemistry 182 Class, Beijing Institute of Petro-Chemical Technology, Beijing, China

[3]Corresponding Author's E-mail: gongliangfa@bipt.edu.cn

Abstract. It is of great importance to investigate the doped TiO2 photocatalyst for the better photocatalytic activity via visible light. In this study, a newly Ag·AgBr@TiO_2 photocatalyst has been prepared via a facile sol-gel process using n-butyl titanate as titanium source, silver nitrate as silver source, and potassium bromide as bromine source. The different nano porous catalyst products were prepared by using of calcining precipitates (obtained by drying the sol-gel in an oven at 80°C) in air at 300, 480, and 660 °C for 3 hours respectively. The selected precursor treatments and heterostructure display significantly improved photodegradation activity on organic 2-naphthol. The microstructures were characterized by XRD and SEM. The results showed that TiO_2 and doped TiO_2 are both anatases generally. Compared with those of undoped TiO_2 and AgBr, the Ag, AgBr-doped TiO_2 has remarkably enhanced photocatalytic activity towards 2-naphthol. Owing to the higher absorption of Ag, AgBr nanoparticles on to the TiO_2, the excellent photocatalytic activities of Ag·AgBr@TiO2 were obtained. The degradation rates of 2-naphthol reached 69% under Ag, AgBr-doped TiO_2 photocatalyst, whereas, with undoped TiO_2 and pure AgBr, the degradation rates of 2-naphthol are 20% and 43% respectively.

1. Introduction

Over the more than two decades, photocatalytic (PC) oxidation technology has growingly been studied for tackling environmental and contaminant problems, as well as energy crisis issues [1-3]. TiO_2 photocatalysts have existed considerable widely research for the dispense of global energy and for dye degradation due to their excellent merits, containing obvious oxidizing performance, non-toxicity, un-solubility in water, and long-term stability, etc. Whereas, bare TiO_2 suffers from the shortcomings (wide bandgap), the utilization of pure TiO_2 for the photocatalytic degradation has been limited significantly. To obtain more remarkable property, it is a promising process that doped TiO_2 by atomic or other photocatalytic materials [4-8]. In this paper, the complex Ag•AgBr@TiO_2 nano-particles have been facily prepared using simple precipitation and photoreduction approach. Subsequently, physicochemical properties of as-fabricated powers were measured by X-ray diffraction (XRD), N_2 adsorption/desorption test, and scanning electron microscope (SEM). The outcomes displayed that the as-fabricated designed photocatalysts existed superior PC performance in transfer and separation of electrons and holes against the degradation of 2-naphthol under the irradiation of the visible light.

2. Experimental

Firstly, under avoiding light, the 250mL aqueous solution containing 0.550 g $AgNO_3$ was added dropwise to the 250mL PEG400 solution (concentration:0.02 mol/L) with KBr (0.52 g), and agitated for 1 hour at room temperature to produce AgBr-PEG. Then, the homogeneous butyl titanate solution (40 mL tetrabutylorthotitanate being dissolved in 80mL of ethanol by stirred for 10 minutes at room temperature) was added to the AgBr-PEG hydrosol at a rate of not more than one drop per second. During the sol-gel process, the mixture was constantly stirred at ambient temperature for an otherwise 2 hours, then aged for 12 hours. Finally, throng many times filtering and washing process using distilled water and pure ethanol, the as-prepared compound was removed water at 80°C in an oven and the obtained precipitate was dried for 3 h to yield white particles, the as-synthesized porous sample was denoted as ABT-T respectively when the calcined temperature as at 300, 480, and 660 °C. But for the combination of AgBr, the bare TiO_2 nano-particles were fabricated by calcining at 500°C for 3 h was labeled with T-500 for comparison.

3. Characterization

X-ray diffraction (XRD) forms of the samples were recorded via a X-ray diffractometer (XRD: PURKINJE GENERAL/XD-2) using graphite monochromator with Cu-K radiation (λ = 0.15406 nm). The increasing voltage and applied current were 40 kV and 40 mA, respectively. The scanning electron microscopy (SEM) was carried out on an Field Emission SEM (QUANTA-400F, FEI, American), and the elevating voltage is 10 KV.

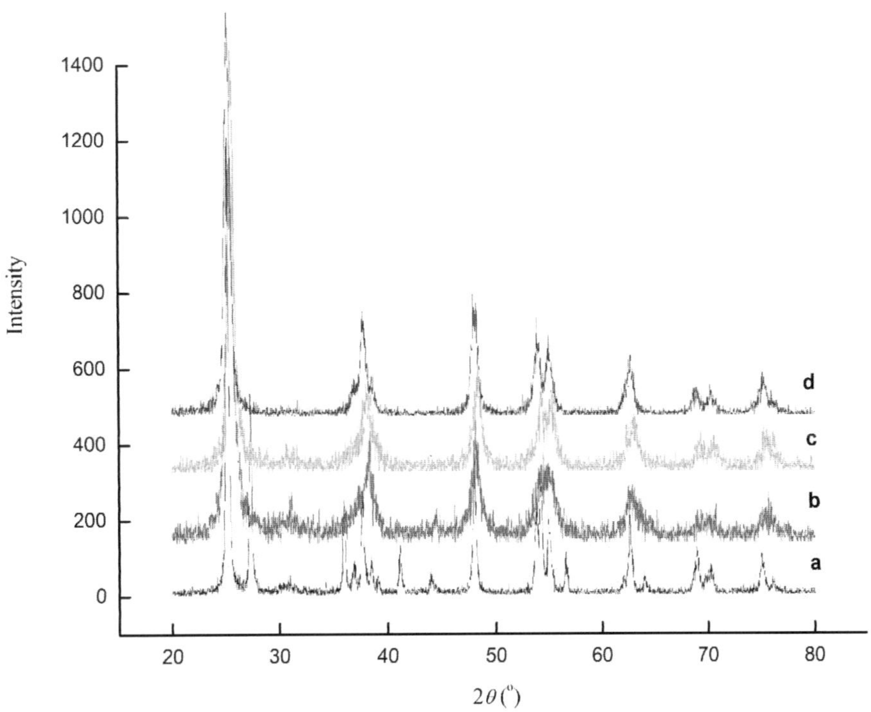

Figure. 1. XRD patterns of (**a**) ABT-300, (**b**) ABT-480, (**c**) ABT-660 and (**d**) T-500.

4. Photocatalytic degradation

The PC activities of the as-fabricated Ag•AgBr@TiO_2 powders were operated under visible light irradiation via the photodecomposition of 2-naphthol (2-NAP) in water solution. The experiment was performed in a photochemical Reactor (DGY-1A, Nanjing Multi-help Technology Development Co.

Ltd. China) equipped with a 150 W Xe arc lamp with a UV cut-off filter (λ > 420 nm), which was as the visible light. In each step, 30 mg catalyst was poured into 50 mL of 2-NAP solution (15 mg L^{-1}) and stirred for 30 min in the dark to obtain the adsorption/desorption equilibrium. At given time intervals, after centrifugation and filtration, the absorption patterns of the centrifuged solutions were operated using a UV-Vis spectrometer (UV-1201, Rayleigh, Beijing), and the concentration of 2-NAP could be obtained from the absorption peak at the characteristic wavelength of 224 nm.

5. Results and discussion

Figure 1 displays the XRD patterns of the samples synthesized at various calcined temperatures. The results show that the Ag•AgBr@TiO_2 prepared by sol-gel method and calcined at 300°C, were composed from the common anatase wtih a tiny amount of rutile phase. Along with the increase of the annealing temperatures, the diffraction peaks of anatase exhibit stronger and sharper. From the full-width at half-maximum of the diffraction peaks, the calculated average crystallite sizes of the ABT-300, ABT-4 80, ABT-660, and T-500 samples are 6.7, 8.1, 11.7 and 21.33 nm, respectively thorough the Scherrer equation. The diffraction peaks at 2θ = 25.2, 48.2, and 62.3° could be ascribed to (101), (200), (204) planes of anatase TiO_2 (JCPDS card NO. 21-1272), while the diffraction peaks at 2θ = 38.1, 44.3, 64.4, and 77.4° could be assigned to (111), (200), (220) and (311) planes of the face-centered-cubic(fcc) silver (JCPDS card NO. 65-2871). The diffraction peaks at 2θ = 30.9, 44.3, 55.1, and 73.2° assigned to (111), (200), (220) and (222) planes of the cubic AgBr (JCPDS card NO. 87-0597) also appeared in figure 1. The XRD patterns also indicate the lower content and smaller particle size of Ag, AgBr and rutile TiO2 phase (JCPDS NO. 21-1276). The incorporation of Ag and AgBr into TiO_2 particles did not have much effects on the crystallites of TiO_2. Figure 2 presents the typical SEM images of the as-fabricated samples. As can be observed that the as-prepared Ag·AgBr@TiO_2 exhibits obvious porous structure with the reduced metallic nanoparticles (NPs) catenation on TiO_2.

Figure 2. Typical SEM images of (**a**) ABT-300, (**b**) ABT-480, (**c**)ABT-660 and (**d**) T-500.

The diameters of silver is largely in the range of 20-60 nm. These microstructures were well deposited on the surface in the whole composite, and which enable the separation properties of photo-generated electron–hole to improve during the photoexcitation process, and enhance photocatalytic efficiency.

Figure 3 displays the photo-decompose nature (conversion C/C_0 (initial C_0 and C are the concentrations at time t = 0 and t = t, respectively)) of as-prepared ABT-480, pure AgBr and pristine TiO_2(T-500) samples against 2-NAP solution. From figure 2, it can be seen that the ABT-300 and ABT-480 samples are analogous each other in the micro-structures, resulting from the well-mixing of AgBr and TiO_2 composite. However, when increase the calcined temperature to 660°C, the core AgBr NPs of Ag•AgBr@TiO_2 photocatalysis begin decomposing into metallic silver, and the influence of plasma resonance on the properties of composite appears, which indicate that the calcined temperature is of importance for the samples structures and photocatalytic efficiency. Thus, ABT-480 has the excellent highest photocatalytic activity against 2-NAP under illumination of visible light (figure 3).

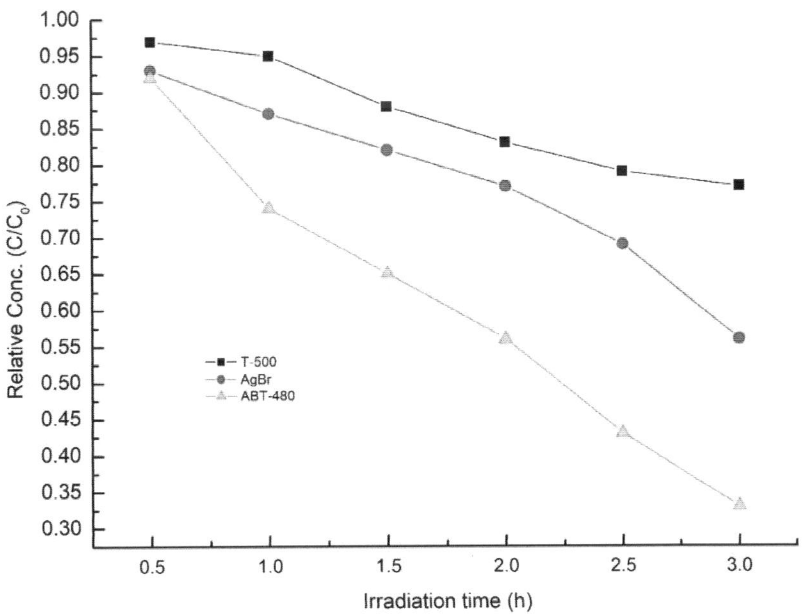

Figure 3. The photocatalyzed degradations of 2-NAP by T-500, pure AgBr and ABT-480.

Comparison with those of pristine TiO_2(T-500) and pure AgBr (figure 3), under which, the 2-NAP solution exhibits very limited degradation, the ABT-460 demonstrate excellent ability for the rapidly decomposed of 2-NAP solution, and for 3 hour of the visible-light illumination, the 2-NAP solutions turn into almost transparent under the ABT-480 action. Thus, it can been say that ABT-480 shows the much higher photocatalytic activity. It is worthy that, the photo-catalytic efficiency of pure AgBr elevated as the light time is longed, owing to the degradation of pure AgBr in the light irradiation, which products metalltic Ag. These NPs can lead to the plasmon resonance effect (PRE).The PRE can enhanced the photocatalytic behavior. This suggests the decreases of recombination efficiency of photo-excited electron–hole pairs.

6. Conclusions

Through sol-gel preparation approach, precipitation and photoreduction processes, a ternary Ag NP deposited Ag•AgBr@TiO_2 photocatalyst was fabricated successfully. The phase structure of the as-synthesized samples was characterized by SEM and XRD. Comparison with the pure TiO_2 and AgBr, the incorporation of ternary Ag, AgBr and TiO_2 improved the separation rate of photo-excited electron and hole, and much more elevating the photodegradation activity against the organic 2-naphthol under visible light illumination. The Ag, AgBr-doped TiO_2 calcined under 480°C has remarkably higher photocatalytic activity for degradation than the undoped TiO_2. This work may lead noval process to prepare the excellent photocatalyst candidates for degradation strategy.

Acknowledgments

This work was supported by the State Education Commission (URT: 2016J00017) funding for undergraduate research and training programs.

References

[1] Chen X B, Shen S H, Guo L J and Mao S S 2010 *Chem. Rev.* **110** 6503-6570
[2] Bohren C F D and Huffman R. 1983 *Absorption and Scattering of Light by Small Particles* (New York: Wiley)
[3] Liang Y H, Lin S L, Liu L, Hu J S and Cui W Q, 2014 *Mater. Res. Bull.* **56** 25-33
[4] Wu D, You H, Jin D and Li X, 2011 *J. Photochem. Photobio. A*, **217** 177-182
[5] Wang Y F and FAN CM. 2012 *Chinese J. Inorg. Chem.* **28** 347-351
[6] Gamage McEvoy J, Cui W and Zhang Z. 2014 *Appl.Cataly. B.* **144** 702-712.
[7] Wang P, Huang BB, Qin XY, Zhang XY, Dai Y, Wei JY and Whangbo M H. 2008 *Angew.Chem. Int. Edit.* **47** 7931-7933
[8] Hu C, Hu XX, Wang LS, Qu J H and Wang A M. 2006 *Environ. Sci. & Techno.* **40** 7903-7907
[9] Umezawa N, Shuxin O and Ye J H. 2011 *J. Phys. Rev. B.* **83** 035202-1-8
[10] A D J R Baltazar 2018 *J. Clust. Sci.* **29** 509-519

Study on template-free preparation of porous g-C_3N_4 and its enhanced visible-light catalytic performance

Chen Li[1], Linkun Xie[1], Ke Xiao[1], Haonan Zhang[1] and Xijuan Chai[1,2]

[1]School of Material Science and Engineering, Southwest Forestry University, Kunming 650224, China

[2]Corresponding author's E-mail: xjchai@126.com

Abstract. Employing thiourea as precursor, the bulk graphitic carbon nitride (named SCN) was prepared via thermal polymerization strategy. The SCN was then stripped and treated by different concentrations of NaOH solution, and then porous carbon nitride SCN with loose structure was obtained and characterized by X-ray diffraction (XRD), N_2 adsorption, scanning electron microscope (SEM), transmission electron microscopy (TEM), photoluminescence (PL). Then catalytic performance of the obtained porous carbon nitride SCN samples was evaluated by degrading methyl blue solution (MB). The results show that the normal pressure alkaline treatment is an effective way to prepare porous carbon nitride. After being treated by NaOH solution of 0.3 mol/L, the specific surface area of SCN increased to three times than that of bulk SCN. The synergistic effect of adsorption and photocatalysis boosts the photocatalytic efficiency of SCN by 1.78 times compared with bulk SCN, and the first-order reaction rate constant K_{app} is increased by 2.9 times.

1. Introduction

In recent years, the shortage of fossil fuel resources and the environmental pollution caused by the rapid industrial development have become critical global issues urging to be resolved [1-3]. Semiconductor photocatalytic technology plays an increasingly important role in degrading hard-to-degrade organic pollutants and advanced treatment of water purification due to its low energy consumption, ease of operation, and environmental friendliness. Among them, graphite phase nitride carbon (g-C_3N_4), a new type of visible photocatalyst obtained attention owing to its unique electronic structure, excellent thermal stability, and chemical stability, as well as its affordability [4-6].

The thermal polymerization method is the main method to prepare g-C_3N_4 in recent years due to its ease of operation [7]. However, most of the g-C_3N_4 prepared by thermal polymerization presents a dense bulk structure, and the layered structure is seriously agglomerated, resulting in a lower specific surface area, greatly weakened photo-generated carrier separation ability, and reduced photoelectric catalytic activity [8-11]. The modification and optimization methods for g-C_3N_4 are mainly through non-metallic, metal doping and co-doping modification [12], designing g-C_3N_4 with different morphologies [13], and optimizing its molecular structure [14]. Among those methods, designing and preparation of porous g-C_3N_4 is beneficial to enhance visible light absorption capacity, expanding the wavelength of light response, increasing the specific surface area, and promoting the effective separation of photogenerated carriers [15-16]. Currently, the template method is still the dominant method for preparing porous g-C_3N_4. However, the template method requires the use of a strong acid or alkali solution to remove the template or the addition of various surfactants, which are not beneficial to environmental protection. Therefore, it is of great significance to explore the template-free preparation of porous g-C_3N_4.

In this study, bulk g-C_3N_4 was prepared using thiourea as a carbon source, and then NaOH solution was used to treat the bulk g-C_3N_4 at atmospheric pressure. The action of alkali is used to reduce the interlayer hydrogen bonding force of bulk g-C_3N_4, and then to destroy the block structure of thermally polymerized g-C_3N_4, thereby obtaining g-C_3N_4 with a large specific surface area. The photodegradation efficiency of g-C_3N_4 treated with different concentrations of alkaline solution is significantly higher than that of bulk g-C_3N_4.

2. Experimental methods

2.1. Sample Preparation

10g of thiourea was weighed in a quartz boat. It was then placed in the middle of a tube furnace and heated at 550 °C for 4h, with a heating rate of 10°C/min. The thiourea was naturally cooled to room temperature after the reaction and milled to obtain massive carbon nitride, which was recorded as SCN. 1g of SCN was taken and placed in four individual 30mL of NaOH solutions with concentrations of 0.1, 0.2, 0.3, and 0.4mol/L, respectively, and stirred at 95°C for 9h. After the reaction, the samples were centrifugated, dried, and grounded into powder. The samples obtained were each marked as SCN0.1, SCN0.2 and SCN0.3, SCN0.4, respectively.

2.2. Characterization

Nova450 Scanning Electron Microscope was used to analyze the morphology of the samples. The sample crystals were measured using a transmission electron microscope (Tccnai G2 TF-30, Hitachi, Japan) to accelerate the voltage by 160kV. The pore size distribution and BET specific surface areas of the samples were analyzed using a gas adsorbent (ASAP-2020, Quantachrome Ins, USA). N_2 was used as the absorbate at a test temperature of 77 K. The light absorption properties of the samples were measured using UV-visible absorption spectroscopy (UV-4100, K-Alpha plus, TMO, USA). The separation efficiency of the sample photogenic carriers was determined at room temperature using photoluminescence spectroscopy (Fluorolog3-21, Hitachi, Japan), and the xenon lamp was the excitation light source (-325nm).

2.3. Photocatalytic testing

Employing 350W Xe lamp light source with a $\lambda >$ 400nm filter as the light source, the photocatalytic performance of the sample was characterized by the efficiency of photodegradation of the MB solution. A sample of 0.1g carbon nitride was dispersed in 100mL of MB solution with a concentration of 10mg/L, stirred with a magnetic stirrer for 30 minutes under no light conditions to reach the equilibrium of adsorption and desorption, and then the xenon lamp was turned on to start the photocatalytic degradation reaction. 3 mL of the suspension for centrifugation was collected every 10 minutes, and the absorbance of the upper liquid was measured to calculate the degradation rate of the sample to MB.

3. Results and discussion

3.1. XRD analysis

XRD images of SCN, SCN0.1, SCN0.2, SCN0.3 were shown in Figure 1. As seen from Figure 1, all samples exhibited diffraction peaks at 13.9° and 27.5°. Thus, it indicated that alkaline treatment did not changed the original crystal structure of carbon nitride. Among them, the diffraction peak strength of 27.5° is the highest, which is the accumulation characteristic peak caused by the conjugate aromatic layer structure, corresponding to the (002) facet of carbon nitride. The diffraction peak at 13.9° is is attributed to the periodic arrangement of the triazine ring in the graphite-like phase layer, corresponding to the (100) facet of carbon nitride. However, it is not difficult to find that with the increase of alkali concentration, the diffraction peaks of the samples at 2θ=27.5° gradually shifts to a higher angle direction. This is because the NaOH solution destroys the interlayer structure and intermolecular hydrogen bonds of SCN, which in turn causes the interlayer spacing of carbon nitride to increase. After alkali treatment, the peak intensity of carbon nitride at 2θ=13.9° is also reduced, indicating that NaOH treatment not only weaken the interlayer force, but also destroys the intralayer network structure of carbon nitride to a certain extent [16].

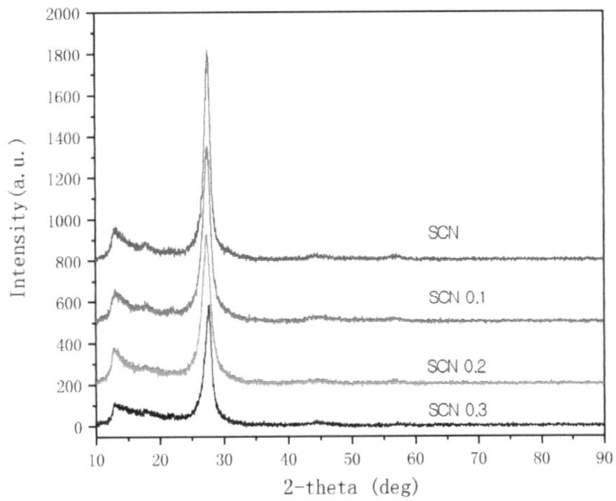

Figure 1. XRD patterns of the prepared samples

3.2. SEM and TEM analysis

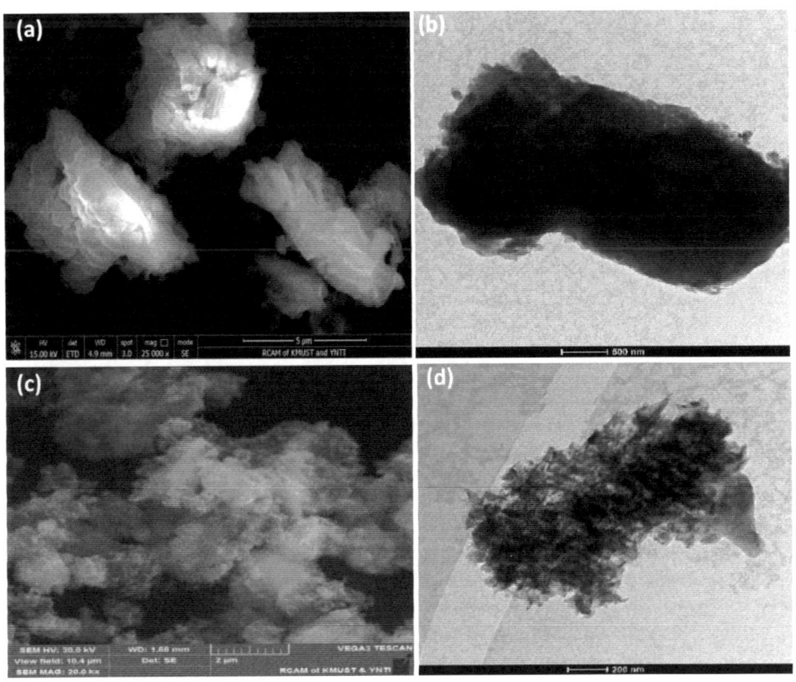

Figure 2. SEM and TEM images of SCN and SCN0.3

Figure 2 (a), (b) are the SEM and TEM images of SCN. As can be seen from Figure 2 (a), SCN shows a whole, large size and obvious layer-to-layer aggregation structure with a smooth surface. Figure 2 (c) and (d) are the SEM and TEM images of SCN0.3, which shows that SCN0.3 exhibit a cotton-like porous structure. This is because NaOH solution destroys the hydrogen bonds between the nitride molecular layers, thus destroying the compact stacked structure of SCN [17]. In the experiment, it was found that with the increase of alkali concentration, the fluffiness of the resulting samples gradually increased. The loose volume and porous structure are more conducive to the adsorption of the organic pollutions, and the photocatalytic efficiency of SCN is improved. Comparing the TEM image of SCN and SCN0.3, the structure of SCN presented a large lamellar structure, while lots of holes can be observed on the chip layers of SCN0.3 with a relative loose structure. The results showed that the samples treated by NaOH solution had a porous structure.

3.3. BET analysis

Figure 3 shows the N_2 adsorption and desorption isotherms of SCN and SCN0.3. Among them, the SCN exhibited the type III adsorption isotherms with a type H3 hysteresis, revealing the presence of abundant slit pores. The N_2 adsorption capacity of carbon nitride at higher pressure increases obviously, which is because that a large amount of N_2 enters the slit pore caused by the accumulation of carbon nitride lamellar structure. The adsorption and desorption curve of SCN0.3 showed an IV type curve and an H2 type hysteresis loop. The H2 type hysteretic loops are generally considered to be caused by the adsorption of porous adsorbents. This indicates that SCN0.3 has a porous structure. It can be seen from Table 1 that SCN has a larger pore size and a smaller pore volume, and its characteristics are consistent with the characteristics of slit holes. In contrast, SCN0.3 exhibited smaller pore size, larger pore volume and higher specific surface area than those of SCN. Its specific surface area of SCN0.3 is 3 times than that of SCN. All in all, the BET characteristics of SCN0.3 further illustrate that the carbon nitride after NaOH solution treatment has a porous structure.

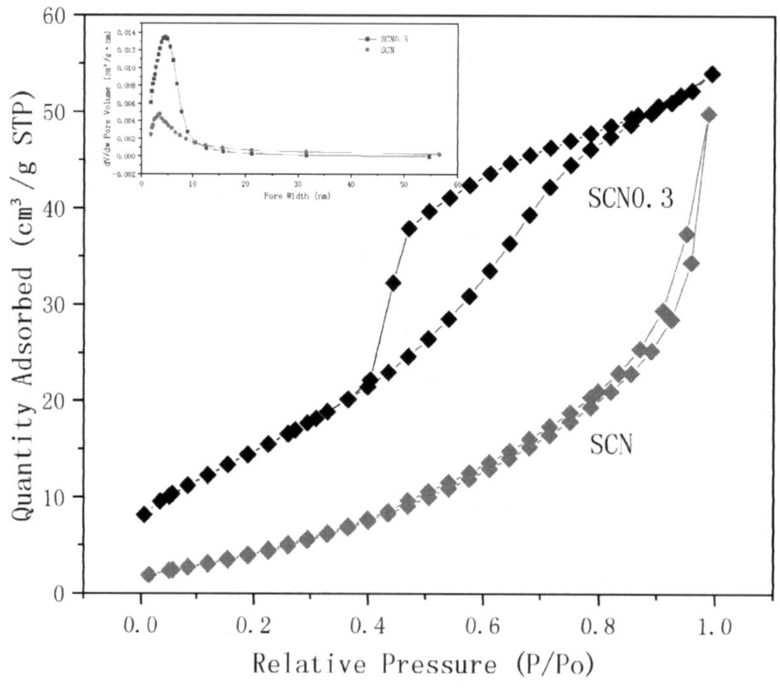

Figure 3. BET images of SCN and SCN0.3

Table 1. Specific surface area and pore structure of SCN and SCN0.3

Sample Name	BET Specific Surface Area (m^2/g)	Mean pore diameter (nm)	Pore Volume (cm^3/g)
SCN	19.01	9.45	0.077
SCN0.3	57.14	3.96	0.086

3.4. PL analysis

The intensity of the PL fluorescence spectrum is related to the recombination efficiency of photo-generated carriers. The higher the recombination rate of photo-generated carriers, the stronger the intensity of the corresponding PL fluorescence spectrum. Figure 4 shows the PL spectra of samples treated with different concentrations of NaOH solutions at an excitation wavelength of 370nm. It can be seen from Figure 5 that the shapes of the fluorescence curves of the samples SCN, SCN0.1 and SCN0.3 are similar, and the emission peaks share similar wavelengths located at 460nm, which is consistent with the previous literatures [8-10]. Comparing the PL intensity of SCN, SCN0.1 and SCN0.3, it is apparent that the PL intensity of the samples was significantly reduced after being treated with NaOH solution. Among them, the emission intensity of the SCN0.3 sample was weaker, which indicated that the photogenerated electrons and holes in SCN0.3 have effectively migrated, and the recombination rate was significantly reduced. This is because, after being treated with NaOH solution, the lamella of SCN0.3 became thinner, which is conductive to greatly accelerating the migration speed of photogenerated carriers from the inner structure of g-C_3N_4 to the surface and then reducing the energy loss in the transfer process. Therefore, a lower PL intensity was exhibited.

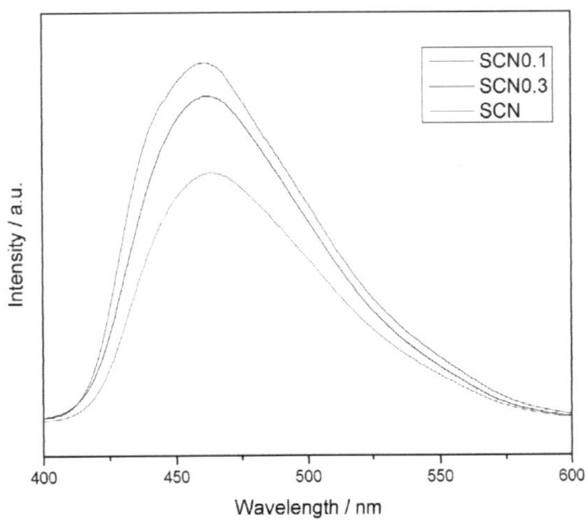

Figure 4. PL spectrum of samples before and after NaOH solution treatment

3.5. Photocatalytic performance analysis

Figure 5 shows the photocatalytic degradation performance curve (a) and degradation kinetics curve (b) of the samples treated by different concentrations of NaOH solution. The figures show that as the concentration of sodium hydroxide increases, the ability of carbon nitride to absorb and photocatalyzed MB degradation gradually increases. When the concentration of NaOH solution was 0.3mol/L, the photocatalytic efficiency of SCN0.3 reached the maximum value of 93.0%, which is 1.78 times to that of SCN. The first-order reaction rate constant K_{app} of SCN0.3 is 2.9 times than that of SCN. The increase in adsorption and photocatalytic degradation is mainly attributed to the thinned porous sheet and increased specific surface area of SCN0.3. The former greatly improves the migration rate and yield of photogenerated carriers from the inside of g-C_3N_4 to the surface. The large specific surface area of SCN0.3 makes it possess more active sites, thus effectively improving the adsorption capacity of g-C_3N_4. It can be seen from the figure that the optimal concentration range of NaOH solution is 0.2mol/L-0.3mol/L.

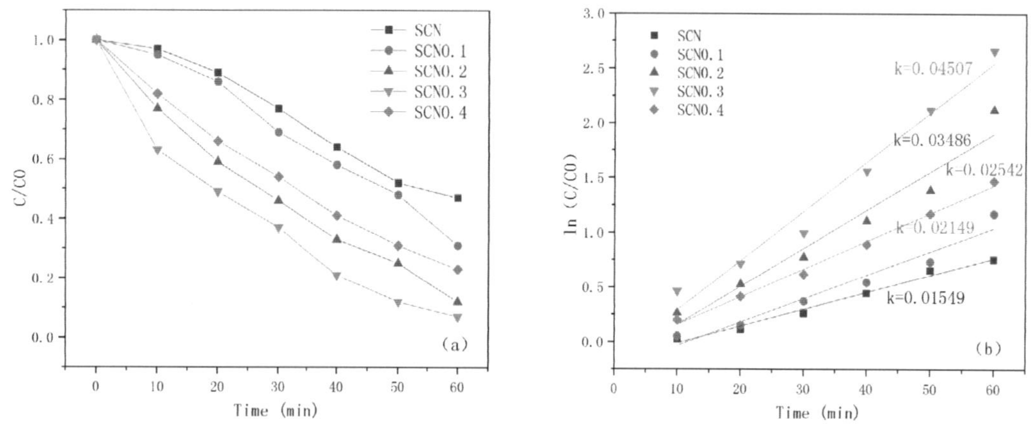

Figure 5. Photocatalytic degradation performance curves (a) and corresponding kinetic curves (b) of all samples

4. Conclusion

Bulk g-C_3N_4 was prepared using thiourea as a carbon source and atmospheric pressure treatment was carried out using NaOH solution. The results of photocatalysis showed that treating in NaOH solution under atmospheric pressure was an effective means to prepare porous carbon nitride with a high ratio of surface area, and the optimal treatment concentration of NaOH is 0.2 mol/L to 0.3mol/L. The loose porous structure and large surface area were the main reasons for the improvement of alkaline treated carbon nitride photocatalytic performance. The results of SEM and TEM analysis showed that the bulk g-C_3N_4 was transformed from aggregated structure to a loose, porous cotton shape after NaOH solution treatment. BET analysis showed that the specific surface area of carbon nitride was greatly increased after sodium hydroxide treatment. The larger specific surface area provided more reactive sites for SCN0.3 and greatly improved the adsorption capacity of g-C_3N_4. The results of PL spectroscopy showed that SCN0.3 had the weakest peak strength of PL emission, which indicated that the compound rate of photogenic carriers in its system was significantly reduced.

Declaration of Competing Interest

The authors declare that they have no known competing financial interests or personal relationships that could have appeared to influence the work reported in this paper.

Acknowledgments

This research was financial supported by the National Natural Science Foundation of China (No. 31960297, No. 31660179, No. 31760184).

References

[1] Fujishima A, Honda K. Electrochemical photolysis of water at a semiconductor electrode [J]. *Nature*, 1972, **238**(5358): 37-38.
[2] Jiang J, Li H, Zhang L. New insight into daylight photocatalysis of AgBr@ Ag: synergistic effect between semiconductor photocatalysis and plasmonic photocatalysis [J]. *Chemistry–A European Journal*, 2012, **18**(20): 6360-6369.
[3] Pelaez M, Nolan N T, Pillai S C, et al. A review on the visible light active titanium dioxide photocatalysts for environmental applications [J]. *Applied Catalysis B: Environmental*, 2012, **125**:

331-349.
[4] Wang X, Maeda K, Thomas A, et al. A metal-free polymeric photocatalyst for hydrogen production from water under visible light [J]. *Nature materials*, 2009, **8**(1): 76-80.
[5] Ding F, Yang D, Tong Z, et al. Graphitic carbon nitride-based nanocomposites as visible-light driven photocatalysts for environmental purification [J]. *Environmental Science: Nano*, 2017, **4**(7): 1455-1469.
[6] Li Y, Ouyang S, Xu H, et al. Targeted exfoliation and reassembly of polymeric carbon nitride for efficient photocatalysis [J]. *Advanced Functional Materials*, 2019, **29**(27): 1901024.
[7] Yao Xiaxi, Zhang Wenjun, Huang Jialei, Du Zhongwei, Hong Xuekun, Chen Xuefeng, Hu Xiuli,Wang Xuhong. Enhanced photocatalytic nitrogen fixation of Ag/B-doped g-C_3N_4 nanosheets by one-step in-situ decomposition-thermal polymerization method [J]. *Applied Catalysis A: General*, 2020 (prepublish).
[8] Nanotechnology - Photocatalytics; Research Results from Guangdong University of Technology Update Understanding of Photocatalytics (Ultrathin Ag_2WO_4-coated P-doped g-C_3N_4 nanosheets with remarkable photocatalytic performance for indomethacin degradation) [J]. *Nanotechnology Weekly*, 2020.
[9] Patnaik Sulagna, Sahoo Dipti Prava, Parida Kulamani. Recent advances in anion doped g-C_3N_4 photocatalysts: A review [J]. *Carbon*, 2021, **172**.
[10] Yaya Wang, Xiao Zhang, Yingjie Liu, Yubin Zhao, Cong Xie, Yuxiang Song, Ping Yang. Crystallinity and phase controlling of g-C_3N_4 /CdS hetrostructures towards high efficient photocatalytic H_2generation [J]. *International Journal of Hydrogen Energy*, 2019, **44**(57).
[11] Deng Y, Tang L, Feng C, et al. Construction of plasmonic Ag and nitrogen-doped graphene quantum dots codecorated ultrathin graphitic carbon nitride nanosheet composites with enhanced photocatalytic activity: full-spectrum response ability and mechanism insight [J]. *ACS applied materials & interfaces*, 2017, **9**(49): 42816-42828.
[12] M. Faisal, Mohammed Jalalah, Farid A. Harraz, Ahmed Mohamed El-Toni, Aslam Khan, M.S. Al -Assiri. Au nanoparticles-doped g-C_3N_4 nanocomposites for enhanced photocatalytic performance under visible light illumination [J]. *Ceramics International*, 2020.
[13] Qidi Zhong, Hanyang Lan, Mengmeng Zhang, Hao Zhu, Ming Bu. Preparation of heterostructure g-C_3N_4/ZnO nanorods for high photocatalytic activity on different pollutants (MB, RhB, Cr(VI) and eosin) [J]. *Ceramics International*, 2020, **46**(8).
[14] Quanhao Shen, Naixu Li, Rehana Bibi, et al. Incorporating nitrogen defects into novel few-layer carbon nitride nanosheets for enhanced photocatalytic H_2 production. *Applied Surface Science*. 2020, **529**:147104
[15]Wendong Zhang, Zaiwang Zhao, Fan Dong, Yuxin Zhang. Solvent-assisted synthesis of porous g-C_3N_4 witheff icient visible-light photocatalytic performance for NO removal [J]. *ScienceDirect*, 2017, **38** (2): 372-378.
[16] Xiaoyu Liu, Lirong Yang, Xingyi Wu. Bul-k g-C_3N_4 exfoliated by NaOH under normal pr-essure and its photocatalytic performance. *Journal of Functional Materials*, 2019, **3**(50),03195-03200.
[17] Qing Zhang, Shuangzhi Liu, Yongcai Zhang, Aiping Zhu, Jing Li, Xihua Du. Enhancement of the photocatalytic activity of g-C_3N_4 via treatment in dilute NaOH aqueous solution [J]. *Materials Letters*, 2016, **171**.
[18] Liang Q, Li Z, Huang Z H, et al. Holey graphitic carbon nitride nanosheets with carbon vacancies for highly improved photocatalytic hydrogen production [J]. *Advanced Functional Materials*, 2015, **25**(44): 6885-6892.

IOP Conf. Series: Materials Science and Engineering 1167 (2021) 012008 doi:10.1088/1757-899X/1167/1/012008

Effect of FSP parameters on microstructure evolution and mechanical properties of AZ31 magnesium alloy

Suna Cha[1], Hongliang Hou[1,3], Yanling Zhang[1], Enyu Guo[2] and Shaojun Wu[1]

[1]Aeronautical Key Laboratory for Plastic Forming Technology, Beijing Key Laboratory of Digital Plasticity Forming Technology and Equipment, AVIC Manufacturing Technology Institute, Beijing 100024, China
[2]Key Laboratory of Solidification Control and Digital Preparation Technology (Liaoning Province), School of Materials Science and Engineering, Dalian University of Technology, Dalian, 116024, China

[3]Corresponding Author's E-mail: hou_hl@163.com

Abstract. To investigate the microstructure evolution and mechanical properties of materials treated by friction stir processing process at various the shoulder sizes and rotating speeds of the tool, a series of friction stir processing experiments were carried out on the AZ31 magnesium alloy with thickness of 4mm by using tool shoulder sizes of 12, 16 and 18mm, and two kinds of rotating speeds of 750r/min and 1500r/min with a constant feed speed of 60mm/min were supplied. Based on the analysis of morphology and mechanical properties of the materials, results demonstrated that reduction of tool shoulder size and rotating speed would contribute to more remarkable the grain refinement effect and affect the strong texture and extreme density strength of the material, leading to higher microhardness. However, lower mechanical properties of yield strength and tensile strength were obtained at these parameters owing to the existence of internal organization defect and stronger defect sensitivity of it. All of the findings would further deepen the understanding of the technique of friction stir processing and contribute to the application of AZ31 magnesium alloy treated by friction stir processing.

1. Introduction

Magnesium and its alloys have the characteristics of good thermal conductivity and electromagnetic shielding, high specific strength, specific stiffness, high damping, excellent casting and cutting performance and thus is widely used in automobile, aerospace, national defense, military industry, electronics and other civil products [1, 2]. However, generally magnesium alloys possess relatively poor plasticity at room temperature because of the limited sliding systems during plastic deformation [3]. Great efforts have been made to improve the plasticity of magnesium alloy by multiple methods such as refining the grains, controlling texture and activating the non-base slip system, etc. [4].

Friction stir processing (FSP), initially proposed by Professor Mishra in 1999, is an important metal processing method that connects metallic components via metallurgical reaction [5]. This technique is developed on the basis of the solid phase connection technology, friction stir welding (FSW) [6]. It is used to soften and maintain high plasticity state of the material by using the friction heat produced by friction between the tool and the workpiece being processed and the stirring extrusion effect on the metal. After the severe plastic deformation, a fine, even and dense microstructure is created within the material. This method is now widely used in the surface modification of materials, refinement of metal microstructure, superplastic materials and the preparation of composite materials [7, 8]. Many scholars have used FSP method to refine the grain size of magnesium alloys, and to improve the uniformity of structure which in turn enhanced the plastic processing ability of magnesium alloys. Yu *et al.* found that the grain size of AZ31 magnesium alloy was decreased from 2.2 μm in air to 0.9 μm after FSP treatment in circulating water [9]. Han *et al.* treated AZ31 magnesium alloy by FSP at the feed speeds of 200 and 400 mm/min, and found that when the rotating speed was less than 1000r/min, the average grain size was less than 7 μm, and the microstructure was coarsened using rotating speed above 1000r/min [10]. However, when Xie treated Mg-Zn-Y-Zr heat-resistant magnesium alloy by FSP, they found that with the increase of rotating speed, the ratio of high angle grain boundary increased greatly, and the grain size was refined to some extent [11]. It can be seen that the FSP parameters in the friction process has a great influence on the microstructure of the material.

However, most of the researches focus on the influence of the process parameters such as rotating speed and feed speed on the FSP. The related reports on the structure of the coupling tool are relatively rare. Therefore, this paper intends to explore the microstructure evolution and resultant mechanical properties of materials treated by FSP process (FSPed materials) at various FSP parameters, especially different the shoulder sizes and rotating speeds of the tool. And then based on the investigations, a more systematic research concerning technique of FSP was reported by us. For this purpose, the AZ31 magnesium alloy with thickness of 4mm was used in the FSP experiments, and the tool shoulder sizes of 12, 16 and 18mm and two kinds of rotating speeds of 750r/min and 1500r/min were supplied while feed speed was remained at 60mm/min. The results demonstrated that reducing tool shoulder size and rotating speed would facilitate more significant the grain refinement effect and the improvement of strong texture and extreme density strength of the material according to micromorphology analysis of microscope and EBSD, leading to higher microhardness. However, lower mechanical strength was obtained at these parameters owing to the existence of internal organization defect and stronger defect sensitivity of it. On the basis of the findings, the understanding of the technique of FSP would be deepened and contribute to the application for FSP in the field of AZ31 magnesium alloy.

2. Experimental materials and methods

In this study, rolled plate of AZ31 magnesium alloy with 4 mm thickness was used as the experimental object to carry out FSP. The composition of AZ31 was listed in Table 1, and the microstructure of the material was shown in Figure 1. The used material has an average grain size of 11.8μm. FSP experiments were carried out using different diameters of the tool shoulder and rotating speeds. The sizes of the tool shoulder used for the FSP experiments were 12, 16 and 18mm. and the rotating speed was set at 750 r/min and 1500 r/min, respectively, while a constant feed speed of 60 mm/min was supplied. The detailed processing parameters were summarized in Table 2.

Each sample for metallographic observations was machined from the processed plate. The cross section of the processing path under each parameter was polished for the microstructure observation. The sample was mechanically polished first using silicon carbide paper, and then polished by W0.5 diamond grinding paste. The surface of sample for microstructure characterization was corroded by picric acid solution (picric acid 5.5 g+ ethanol 90 ml + acetic acid 5 ml + distilled water 10 ml). The microstructure of the samples was observed by scanning electron microscope (Leica DMI5000M) and optical microscope (Zeiss supra55， Germany). Electron backscatter diffraction (EBSD) analysis was performed to evaluate the changes of texture orientation, and a step size of 1μm was used for the EBSD measurements. The mechanical properties of the samples were measured by Vickers hardness and tensile properties. The hardness of the samples was measured by full-automatic Vickers hardness tester (tukon2500-6), tensile strength and yield strength were detected by universal mechanical testing machine.

Table 1. Chemical composition of AZ31 magnesium alloy (wt.%)

Al	Zn	Mn	Si	Fe	Cu	Ni	Mg
2.5~3.5	0.6~1.4	0.2~1.0	<0.08	<0.003	<0.01	<0.001	Bal.

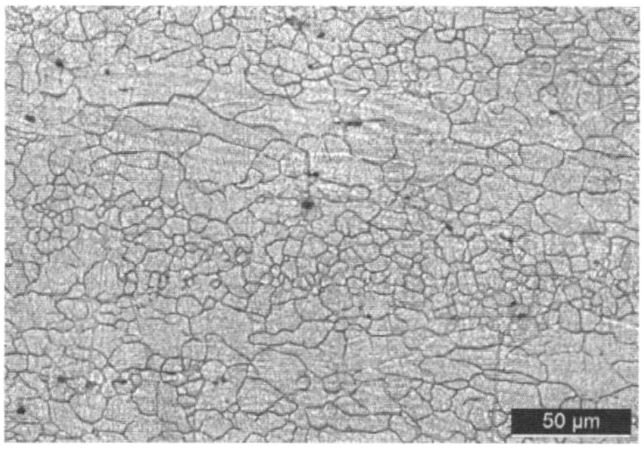

Figure 1. Microstructure of AZ31 magnesium alloy

Table 2. Processing parameter list

Specimen numbers	1	2	3	4	5	6
shoulder diameters/ rotation speed	φ 12mm/ 750r/min	φ 12mm/ 1500r/min	φ 16mm/ 750r/min	φ 16mm/ 1500r/min	φ 18mm/ 750r/min	φ 18mm/ 1500r/min

3. Results and analysis

3.1. Analysis of morphology and grain size

The macrostructure of materials after FSP with various processing parameters is shown in Figure 2. At the rotating speed of 750r/min, when the diameter of tool shoulder was 12mm, there exist many large edge fins on the processing surface; when the diameter of tool shoulder was 16mm, the edge fins were improved, and there were burrs around it; when the diameter of tool shoulder increased to 18mm, the burr would be further reduced. While the rotating speed was up to1500r/min at every diameter of tool shoulder, the processing areas were smooth than 750r/min. And there were a few burrs when diameter of tool shoulder was 12 and 16mm, the processing area was clean and smooth when the diameter of tool shoulder was 18mm. On the other hand, At the same rotating speed, the heat generated by the friction between the tool shoulder and the materials increased with the increment of diameter of tool shoulder, strengthening the driving force of plastic deformation and compacting the processing interface. And smoother and tidier processing area can be obtained while faster rotation speed and higher heat input were provided at the same diameter of tool shoulder. Figure 3(g) shows the microstructure of the base material (BM) before FSP under various parameters, the BM had different grain sizes and shapes, changing from a few μm to several tens of μm. And Large grains were mostly elongated along the rolling direction and small grains were mostly equiaxed. After FSP the grains in stir zone were more uniform than unprocessed, and all of them tended to be equiaxed. The quantified grain size before and after friction stir processing is summarized in Figure 4. The results revealed that the crystal grains were refined after the material was subjected to FSP, and the crystal grains would be finer while the tool shoulder diameter is reduced under the same rotating speed or reduction of the rotation speed is supplied under same tool shoulder diameter. Additionally, when the tool shoulder diameter was 12mm, the grain refinement was most remarkable at the rotating speed at 750r/min, and the average grain size was obviously reduced by ~62% from 11.8μm of the base material to 4.5μm. The main cause for the obtained decreased grains in materials was that reduction of tool shoulder diameter and the rotating speed would directly decrease the heat input, leading to the suppression of grains growth in the process of FSP.

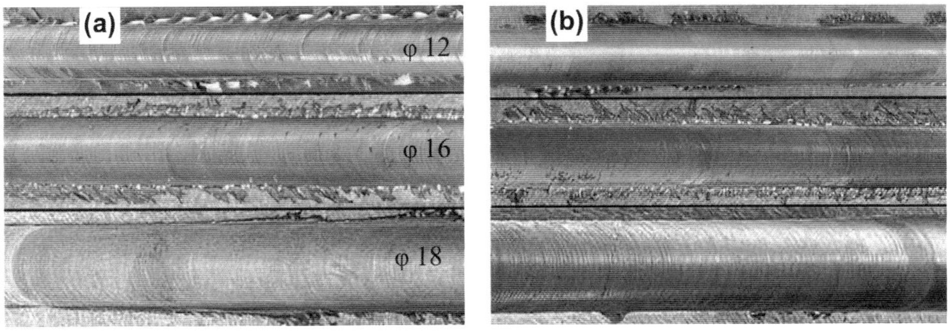

Figure 2. Macro morphology of FSPed AZ31 magnesium alloy with different diameters of tool shoulders and rotating speeds: (a) 750r/min; (b) 1500r/min

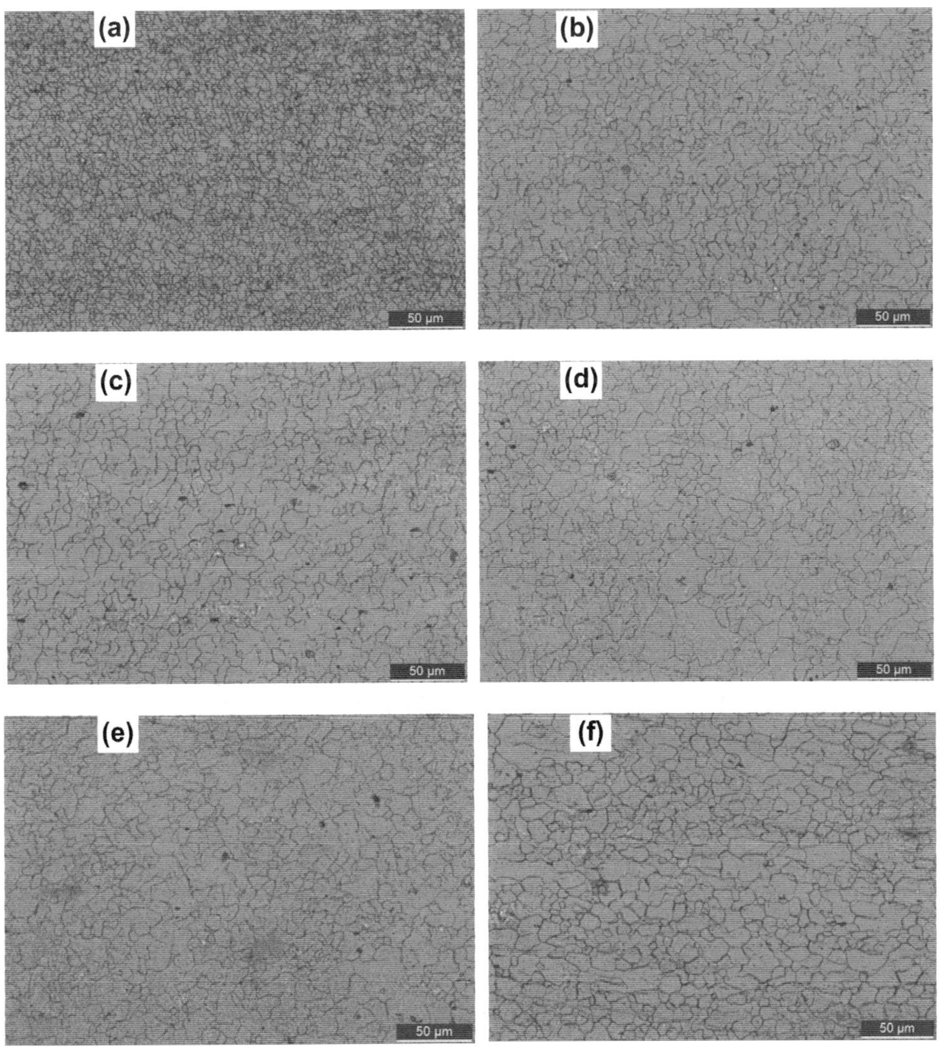

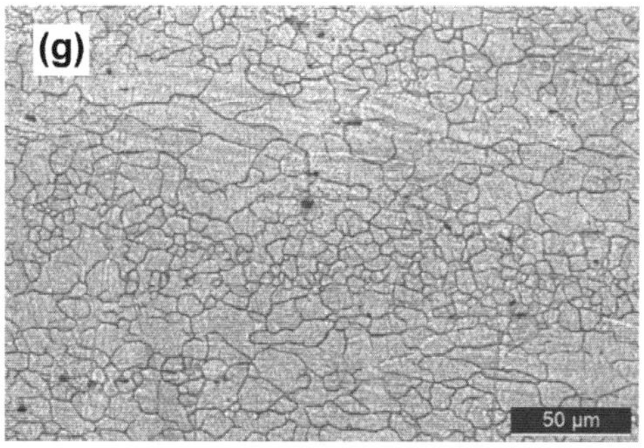

Figure 3. Microstructure of BM and FSPed AZ31 magnesium alloy in stir zone with different diameters of tool shoulders and rotating speeds: (a) φ 12,750r/min; (b) φ 12,1500r/min; (c) φ 16,750r/min; (d) φ 16,1500r/min; (e) φ 18,750r/min; (f) φ 18,1500r/min; (g) BM

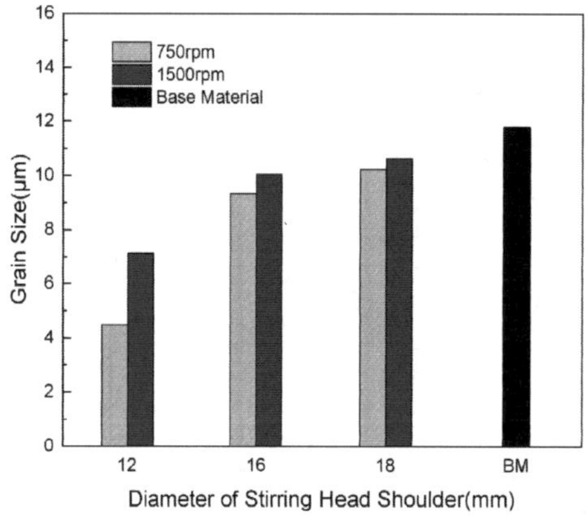

Figure 4. Average grain size of BM and FSPed AZ31 magnesium alloy in stir zone with different diameters of tool shoulders and rotating speeds

3.2. EBSD analysis

To further investigate the grain morphology of the stir zone, EBSD experiment was carried out subsequently. Figure 5 shows the grain orientation and size before and after FSP. It can be seen that the grain orientation after FSP at φ 12-750r/min exhibited significant changes from the original (2110) to the (0001) base surface, which is the sliding surface of the magnesium alloy AZ31. Therefore, the material after FSP was more prone to dislocation slip and plasticity was improved compared with the base material. On the other hand, the grain refinement was received significantly after FSP process with φ 12, 750r/min compared with BM. According to the results of number friction of grain boundary at various angles before and after FSP shown in Figure 6, the small-angle grain boundaries ($<15°$) of the base material accounted for 16.2% while the small-angle grain boundaries after FSP increased by 8.9%. The relationship between the grain boundary energy γ and the difference in orientation conforms to the formula $\gamma = \gamma_0\theta(A - \ln\theta)$, where γ_0 and A are constants, and θ is the difference in orientation. It can be seen that the interface energy of the small-angle grain boundary is lower. So, after FSP the grain boundary energy was lower than that of the base material, contributing to stabler interface. The occurrence of grain refinement and orientation improvement would make an influence on the mechanical properties of the material treated by FSP process, the detailed analysis would be discussed in the following section. Figure 7 present reverse pole images, which demonstrated significant changes of the texture of the materials before and after FSP. The BM had a strong base surface texture on the (0001) plane along the <100> direction, and the extreme density strength was 12.951. However, the material displayed strong texture on the (0001) plane along the <001> direction and (2110) plane along the <010> direction after FSP, and the extreme density strength could reach up to 23.008, which demonstrated that reducing the tool shoulder size and rotating speed facilitated the grain orientation in the stir zone.

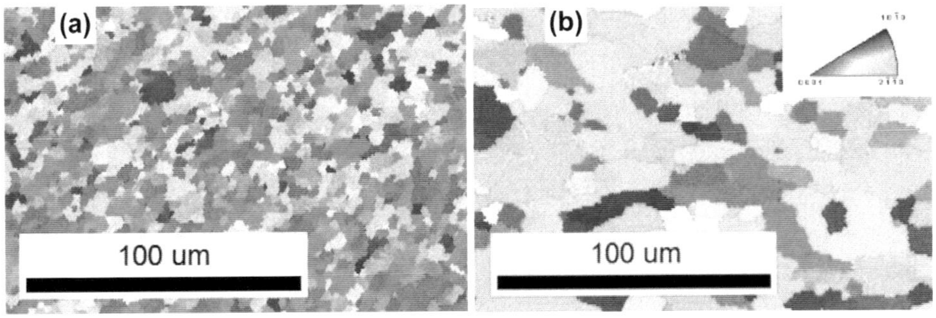

Figure 5. Grain orientation of (a) FSPed AZ31 magnesium alloy in stir zone at φ 12, 750r/min and (b) BM

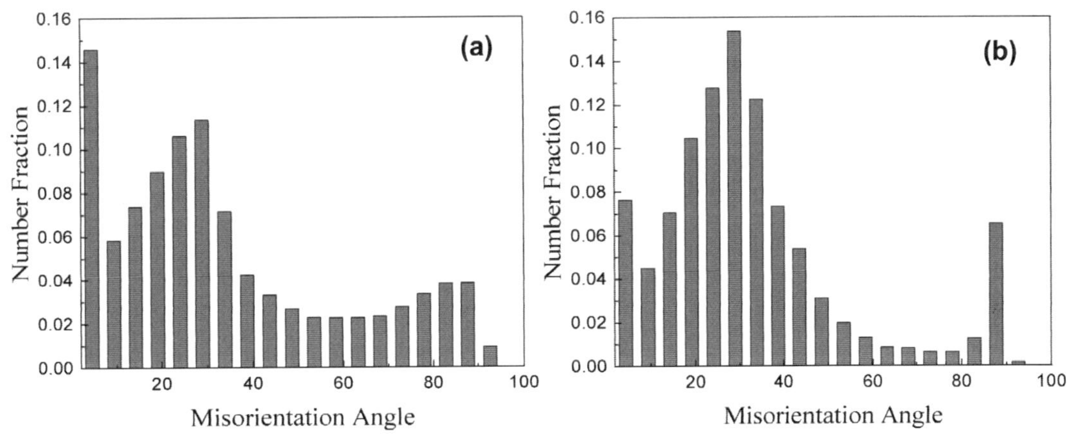

Figure 6. Number friction of grain boundaries of (a) FSPed AZ31 magnesium alloy in stir zone at φ 12, 750r/min and (b) BM

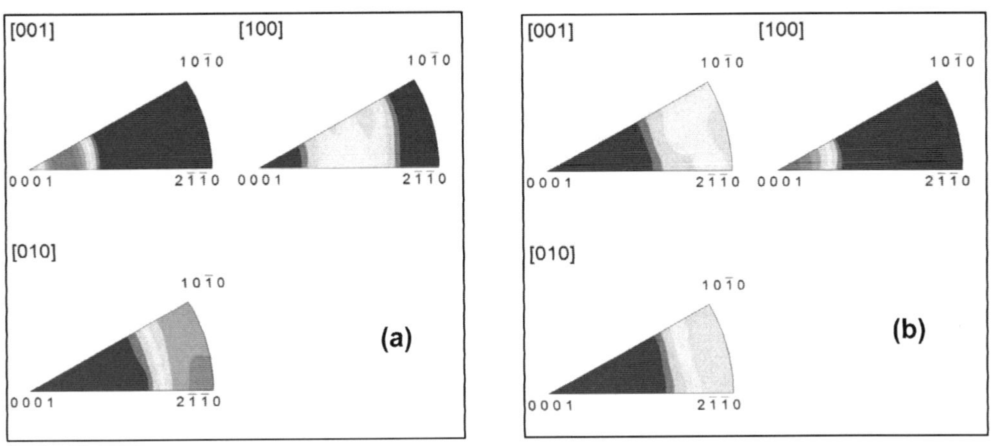

Figure 7. Inverse pole images of (a) FSPed AZ31 magnesium alloy in stir zone at φ 12, 750r/min and (b) BM

3.3. Mechanical properties

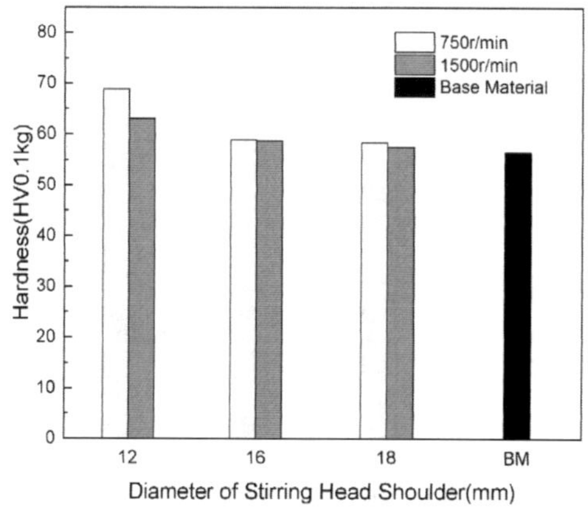

Figure 8. Average hardness of BM and FSPed AZ31 magnesium alloy in stir zone with different diameters of tool shoulders

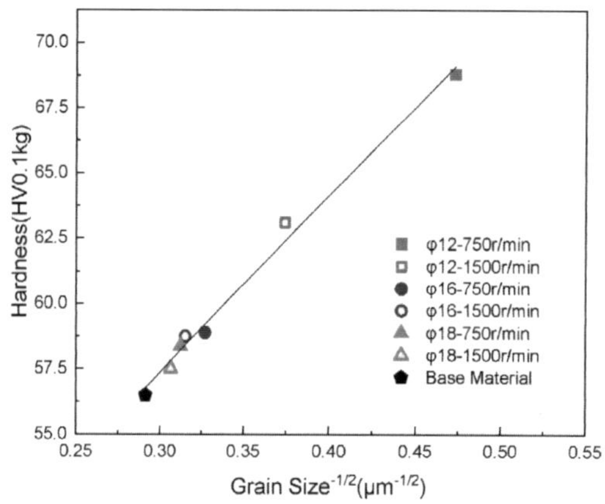

Figure 9. Relation between average grain size and hardness of BM and FSPed AZ31 magnesium alloy in stir zone with different FSP parameters

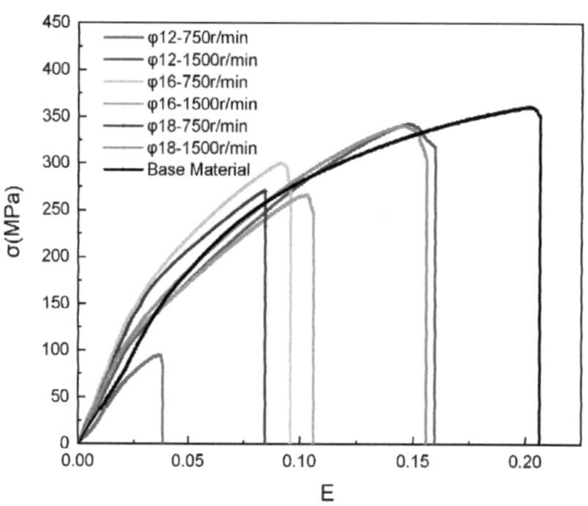

Figure 10. True stress-strain curves of BM and FSPed AZ31 magnesium alloy by uniaxial tensile test at room temperature with different FSP parameters

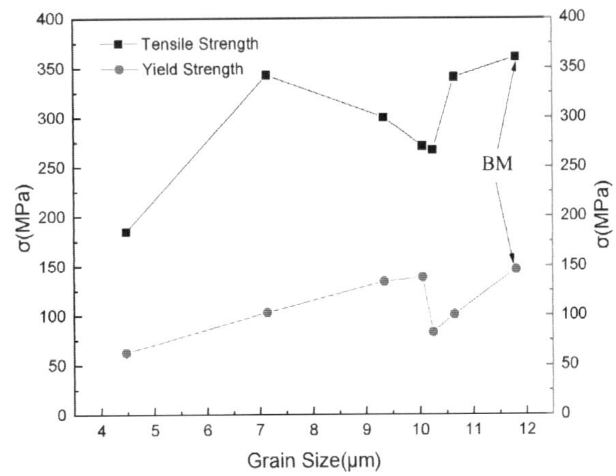

Figure 11. Tensile strength and yield strength of BM and FSPed AZ31 magnesium alloy with different grain sizes by uniaxial tensile test at room temperature

In order to determine the effect of FSP parameters on the mechanical properties of the material, micro-hardness and room temperature tensile test were performed as well. The average hardness value of the stir zone was shown in Figure 8 after FSP with different diameters of tool shoulders at rotating speeds of 750 and 1500r/min. The microhardness of the materials in stir zone after FSP was higher than that of the base material, and reduction of diameter of tool shoulder and rotating speed can contribute to higher hardness value. Especially the highest microhardness of 68.8Hv can be received at rotating speed of 750r/min when the diameter of tool shoulder was 12mm, which was related to average grain size. The average grain size and the microhardness value are plotted in the same figure to further analyze the relationship between the two of them, and the result is shown in Figure 9. It is found that the hardness follows the equation, $H_v = 37.06 + 67.73d^{-1/2}$, revealing that a higher hardness value would be achieved while reduction of grain size of the material was obtained. The relationship follows the Hall-Petch formula, and the results was in accordance with the above morphology analysis of microscope and EBSD. Figure 10 showed the true stress-strain curve after uniaxial room temperature tensile test, consisting of the elastic deformation stage and plastic strengthening stage during stretching process of the material. The flow stress increased linearly with the increase of strain at the elastic deformation stage. Subsequently, the material would continue to elongate with continuously increasing the flow stress, and fracture of the material occurred when the flow stress reached the peak value of tensile strength. Figure 11 illustrated the yield strength and tensile strength under each parameter, the yield strength and tensile strength of the treated materials were lower than that of BM, and there was no obvious linear relationship between strength and grain size. However, the lowest yield strength and tensile strength would be received while the material was treated by FSP at φ12mm-750r/min, along with the earliest fracture. This phenomenon can be explainable by the incompact macroscopic morphology of the processing area and the defects appearing in the fracture zone of the tensile specimen due to lower heat input and incomplete plastic flow during processing, tensile strength and yield strength present higher sensitive to defect, contributing to the observation of the premature fracture of the material.

4. Conclusion

In this paper, investigations of the microstructure evolution and mechanical properties of FSPed materials at various the shoulder sizes and rotating speeds of the tool were carried out, and the morphology and mechanical properties were analyzed by microscope, EBSD, Vickers hardness tester and universal mechanical testing machine. The research revealed that FSP process with various the tool shoulder sizes and rotating speeds produced different variations of the morphology and mechanical properties of AZ31 magnesium alloy. All of the findings would further deepen the understanding of the technique of FSP and contribute to the application of FSPed AZ31 magnesium alloy. The detailed results were list as follow:

1. Reduction of the tool shoulder size and the rotating speed would contribute to more remarkable the grain refinement effect. And when the tool shoulder diameter was 12mm, the most remarkable grain refinement was received at the rotating speed at 750r/min, and the average grain size was obviously reduced from 11.8μm of the base material to 4.5μm. The main cause for the obtained decreased grains in materials was that reduction of tool shoulder diameter and the rotating speed would directly decrease the heat input, leading to the suppression of grains growth in the process of FSP.

2. After the FSP process with a tool shoulder of 12 mm diameter, the proportion of small-angle grain boundaries increased by 8.9% compared with BM, and the material displayed strong texture on the (0001) plane along the <001> direction and (2110) plane along the <010> direction after FSP, the extreme density strength could reach 23.008, while the BM had a strong base surface texture on the (0001) plane along the <100> direction, and the extreme density strength was 12.951.

3. The relationship between the microhardness of stir zone and the average grain size was obtained, which was in accordance with the Hall-Petch relation curves. And as the average grain size decreased, the microhardness increased. While the tensile strength and yield strength of the material were lower than that of BM after grain refinement by FSP, which were due to the existence of internal organization defect and stronger defect sensitivity of mechanical properties.

References

[1] Dong X, Feng L, Wang S, Eric A Nyberg, Ji S. 2021, A new die-cast magnesium alloy for applications at higher elevated temperatures of 200-300 °C. *J. Magnes. Alloy*, **9**(1): 90-101.

[2] Zhang G, Qin S, Yan L, et al. 2021, Simultaneous improvement of electromagnetic shielding effectiveness and corrosion resistance in magnesium alloys by electropulsing. *Mater. Charact*, **174**: 111042.

[3] Wang L, Li Y, Zhang H, et al. 2020, Review: Achieving enhanced plasticity of magnesium alloys below recrystallization temperature through various texture control methods. *J. Mater. Res. Technol*, **9**(6): 12604-25.

[4] Maryam, Jamalian, David P Field. 2020, Gradient microstructure and enhanced mechanical performance of magnesium alloy by severe impact loading. *J. Mater. Sci. Technol*, **36**(01): 47-51.

[5] Mishra R S, Mahoney M W, McFadden S X, et al. 1999, High strain rate superplasticity in a friction stir processed 7075Al alloy. *Scr. Mater*, **42**(2): 163-8.

[6] Dfob A, Se B, Pmgpm B. 2021, Friction stir welding. *Advanced Joining Processes*, 173-206.

[7] Chen Y, Hu H, Huang C. 2018, *Friction Stir Welding Technology*. Beijing: Machinery Industry Press, 1-11. (In Chinese)

[8] Zhu Y, Zuo L, Gu S, Zuo D. 2018, Research Progress in Friction Stir Processing. *Surf. technol*, **47** (4): 221-9. (In Chinese)

[9] Yu L, Zhang Z, Wang K, Wang W, Jia S. 2018, Effect of Friction Stir Processing on Microstructure and Mechanical Properties of AZ31 Magnesium Alloy. *Mater. Rep*, **32**(8): 1289-93. (In Chinese)

[10] Han R, Sha G, Xu R, Wang J, Zhao Z. 2016, Effect of Friction Stir Processing on Microstructure and Tensile Properties of AZ31 Mg Alloy. *Hot. Working. Technol*, **45**(2): 56-9. (In Chinese)

[11] Xie g, Ma Z, Xue P, Luo Z, Wang G. 2018, Effects of Tool Rotation Rates on Superplastic Deformation Behavior of Friction Stir Processed Mg-Zn-Y-Zr Alloy. *Acta. Metall. Sin*, **54**(12): 1745-55. (In Chinese)

A bibliometric analysis of research on acoustic emission for nondestructive testing

Yu Yan[1], Zhensheng Yang[1,3], Haoda Li[1] and Chenming Jiang[2]

[1]School of Logistics Engineering, Shanghai Maritime University, Shanghai 201306, People's Republic of China
[2]China Institute of FTZ Supply Chain, Shanghai Maritime University, Shanghai 201306, People's Republic of China

[3]Corresponding Author's E-mail: yangzs@shmtu.edu.cn

Abstract. Acoustic emission (AE) testing has been an active area of research globally for many decades. The present work is a comprehensive overview based on bibliometric analysis to investigate the application of AE research. The publication trends, key areas, organizations, countries, and sources and publications of AE applications are presented to address the following basic questions: what are the publication sources and trends, what are the key areas, and which countries and organizations are active in AE research? The influences of various AE subjects and topics are also discussed. With the help of the Web of Science Core Collection database and the specialized bibliometric analysis tool VOSviewer, the results are visualized properly, and it is found that AE research is currently developing rapidly.

1. Introduction

Acoustic emission (AE) is a class of phenomena whereby transient stress and/or displacement waves are generated by rapid releases of energy from either localized sources within a material or the transient waves so generated [1]. These rapid energy releases give rise to elastic waves in the ultrasonic frequency regime that can be detected by transducers attached to the surface of the specimen, provided there are waves of sufficient amplitude [2]. In general, the term AE is used to describe both a technique and the phenomenon upon which the technique is based.

The most significant advantage of AE is that it can be used to observe damage processes in materials during the entire load history without disturbing the specimen. Having such a non-destructive testing (NDT) method that can detect and locate flaws in real time is very important. Interest in AE has grown hugely over the past 50 years. Many excellent results have been reported on successful AE applications in engineering for abroad variety of materials, material compositions, and structures. There have also been many research reviews on various aspects of AE.

Although review articles do not provide new methodologies or applications, they contribute significantly by providing a concise up-to-date overview of the development of a field. Because of its superior signal-to-noise ratio and sensitivity at the precision scale, AE has been applied widely in both conventional and precision machining processes. Lee et al. [3] discussed the suitability of AE as a monitoring technique at the precision scale. They found that AE tended to propagate at frequencies well above the characteristic frequencies attributed to machining or natural structural modes in the machine tool structural loop, thereby minimizing the introduction of noise into the AE signal from the phenomenon being studied. These authors demonstrated the sensitivity of AE in three different manufacturing regimes, namely the (i) normal or conventional, (ii) precision, and (iii) ultraprecision scales. Li [4] reviewed the research on AE-based tool wear monitoring in turning and concluded that AE was the area of most-intense research activity for developing intelligent tool condition systems. He suggested that careful signal processing and feature integration with other sensors would be an effective approach for AE-based tool condition monitoring. Kishawy et al. [5] provided a review study to discuss the research into using AE to monitor and control various machining processes, e.g., cutting, turning, and drilling. Their review covered topics including AE processing and analysis methods, hip formation and morphology, surface quality, and tool wear evolution. Jayakumar et al. [6] reviewed the application of AE techniques in grinding, forging, drawing, punch pressing, and blanking.

The application of AE in structural health monitoring (SHM) is another highly active research topic. Ono [7] presented a comprehensive review of AE attenuation and damping in SHM. He focused on the signal loss in large-scale AE monitoring as well as source location, bridge monitoring, sensing and signal processing, and pressure vessels and tanks. He also reviewed the mechanisms and modelling of AE [8] and the application of AE in structural diagnosis [9] and composites [10]. Shiotani [11] and Hay [12] reviewed various applications of AE to infrastructures. Manthei and Plenkers [13] presented state-of-the-art applications of AE for SHM in mines, as well as a comprehensive review of the characteristics and applications of in situ AE monitoring in salt mines and gold mines. Sengupta et al. [14] reviewed the literature on structural damage localization by AE and discussed source localization methods including AE signal processing, applications of artificial neural networks, and finite-element analysis. Holford et al. [15] introduced the challenges faced in developing an AE-based SHM system and reviewed the progress made in addressing these challenges. Nair and Cai [16] presented a brief review of the AE technique and its applications to bridge health monitoring, and they discussed the available techniques for AE data processing, both qualitative and quantitative. Carpinteri [17] proposed an event counting method to determine the AE source in the bending experiment of reinforced concrete beams, effectively reducing the risk of discrete crack diffusion.. Sagar and Prasad [18] reviewed the developments in parametric-based AE signal processing methods applied to concrete structures and presented an overview of the parametric features developed in the assessment of concrete structures. Carpinteri et al. [19] researched a particular methodology by AE technique to evaluate the damage level and final collapse time in reinforced concreted structures. Kurz et al. [20] presented an approach, Akaike Information Criterion (AIC), to automatically pick the onset of ultrasound signals which are used for concrete monitoring and compared it with Hinkley criterion. Noorsuhada [21] carried out a comprehensive review of fatigue damage assessment of reinforced concrete structures with the aid of AE, including the fatigue of reinforced concrete structures, the configuration of the fatigue test, the effect of the fatigue amplitude of the reinforced concrete structure, and the correlation between the AE technique and the fatigue damage of the reinforced concrete structure. This review also highlighted two research gaps in this field.

AE has also been applied successfully in material testing. Tsangouri and Aggelis [22] reviewed the application of AE to concrete and cementitious composites. They provided a comprehensive review of recent studies showing that delicate characterization of the quality of reinforced cementitious media is possible before reaching the dominant damage state, and they concluded that AE can result in early low-energy signals from microscopic displacements. Ahmed et al. [23] and Drummonda et al. [24] reviewed the state of the art in the detection and location of wire breaks through AE monitoring. Rosa et al. [25] reviewed AE applications in studies of natural fibre composites, covering interface studies in single-fiber composite tests, damage evolution, failure-mechanism detection, and crack propagation. Behnia et al. [26] reviewed AE-based methods for assessing the low-temperature cracking performance of asphalt binders and asphalt pavement materials. They presented a brief overview of using AE to estimate the low-temperature performance of unaged and short-term and long-term aged binders, as well as asphalt materials. Romhány et al. [27] reviewed different location methods along with AE characteristics in the time and frequency domains in loaded polymer composites. They discussed the challenge of applying AE, namely establishing a reliable real-time SHM system. Faisal et al. [28] summarized the state of knowledge on AE techniques applied to material-property evaluation during indentation testing. They focused on the phenomena that occur during indentation and how examining them could be used to study more fundamental material behaviour such as deformation and fracture. Michlmayr et al. [29] conducted a comprehensive review of studies that report the use of AE for monitoring failure in geologic materials, and they provided a thorough description of AE generating source mechanisms in mechanically stressed geologic media, measurement principles, and available AE sensors.

To give a comprehensive insight into its potential application to brittle materials, Nejati et al. [30] presented an overview of using AE as a monitoring technique, and they reviewed properties including crack growth behaviour, localization, damage evolution, dynamic character, and structural monitoring. R′ ıos-Soberanis [31] reviewed the application of AE in detecting and identifying initiation sites, crack propagation, and fracture mechanisms of polymer matrix composite and ceramic materials. They highlighted the excellent results of the AE technique as a unique tool for characterizing mechanical parameters in response to external stresses and degradation processes.

Regarding the monitoring of mechanical process, Al-obaidi et al. [32] conducted a brief review on the applications of AE to rotating and reciprocating machinery. Caesarendra et al. [33] presented a literature review on the application of AE to typical rolling element bearings running at high speed, low speed, and very low speed. They focused on AE signal processing, feature extraction, and pattern recognition. Mylvaganam [34] provided a brief overview of AE-based particle monitoring and focused in particular on the monitoring of particle flow. Tan and Mba [35] and Toutountzakis et al. [36] reviewed the developments in using AE to diagnose gear defects and discussed the limitations of AE for identifying seeded defects in gearboxes. Yan et al. [37] reviewed the applications of AE for valve detection, covering condition monitoring, failure, cavitation detection, and the development of portable measuring devices for valves. Yan and Jones [38] reviewed AE testing of large structures such as pressure vessels, storage tanks, pipelines, bridges, and large low-speed bearings. Boyd and Verley [39] reviewed the uses of AE in chemical engineering processes such as gas–liquid mixing, solid systems, and chemical reactions. Gholizadeh et al. [40] reviewed major research developments in applying AE in manufacturing, civil, aerospace, and material engineering. Jemielniak [41], Terchi and Au [42], and Zhao et al. [43] studied AE signal preprocessing and processing methods to provide generic nonparametric methods that assume minimal a priori knowledge about the nature of the signal.

The use of AE in orthopaedics and surgery has risen rapidly in recent years. Kapur [44], Rashid and Pullin [45], and Browne et al. [46] examined the literature surrounding the use of AE in orthopaedics and provided a comprehensive overview of its applications within orthopedic surgery. Olorunlambe and Dearn [47] provided an overview of the tribological and biotribological applications of AE and discussed the viability of using AE for condition monitoring of artificial joints through monitoring their tribological phenomena.

Although the above review papers provide researchers with significant information about AE, there is a distinct lack of review papers that show numerical growth, contribution and effectiveness of involved countries, emerging countries, potential sources of publications, and prevalent factors behind the higher number of citations of a document. These important research parameters will both help and interest researchers worldwide, especially in giving early-career researchers in this field a head start by providing them with a brief overview of the present status of AE research as well as its evolution. To the best of our knowledge, only one existing review paper provides a brief bibliometric analysis of AE methods from 2012 to 2017 [48]. However, that review concentrated on acoustic analysis in quality evaluation and did not cover most of the aforementioned research parameters such as numerical growth and contribution of involved countries. The present work focuses on these important research parameters by means of bibliometric analysis, which is suited to describing such parameters and so is becoming increasingly popular among researchers [49]. The present paper should also appeal to those in industry and encourage more collaborative research between academia and industry.

The remainder of this paper is structured as follows. Section 2 describes the materials and methods in detail. Section 3 presents the results of analysing publications trends, key areas, organizations, countries, and sources and/or publications. Section 4 contains our conclusions.

2. Materials and methods

2.1. Data source

As an important tool for global analysis and investigation in various scientific areas, bibliometric analysis is an approach of document analysis that takes advantage of bibliometric theory to analyze relevant literature using mathematical and statistical methods [50]. Typically, bibliometric analyses are carried out using one of the four widely popular databases of Web of Science (WoS), Scopus, Google Scholar, and PubMed [51]. The WoS is a world leader among citation databases and is produced by Clarivate Analytics. The WoS Core Collection database comprises the Science Citation Index Expanded (SCI-E), the Social Science Citation Index (SSCI), and the Conference Proceedings Citation Index-Science (CPCI-S), among others. The dataset used in the present study was the WoS Core Collection database, which is regarded as the most significant and frequently used scientific database in many research fields.

The search fields of the WoS Core Collection comprise topic, title, author, publication name, year published, address, and author identifiers. Note that the topic field actually contains four fields, namely title, abstract, author keywords, and Keywords Plus. The latter is produced by Clarivate Analytics to highlight additional relevant but overlooked keywords that were not listed by the author or publisher.

2.2. Search strategy

To exclude documents in which AE is mentioned in the abstract only and is not the research topic (e.g., in [52], the research topic is transformer saturation in high-power three-phase converters; AE is mentioned only once in the abstract and therefore should not be considered as a research focus), we restricted the search items to the "title" field only. However, to avoid omissions, we searched for AE OR "acoustic emission*" so that as many documents as possible related to AE were covered. We used the asterisk (*) in the search as a wildcard character to make our search simpler and more comprehensive by tracking all possible forms of the used terms. However, because AE can have meanings other than "acoustic emission" (e.g., AE can mean "associate editor" or "adverse event"), the search results were checked item by item to identify irrelevant keywords and remove them using the Boolean operator NOT in the topic field. The Boolean operators OR, AND, and NOT were used to construct the search strings for evaluating the total number of publications related to AE.

For publications up to February 28, 2020, this search resulted in the bibliographic details of 9410 documents considering only articles (A), proceedings papers, reviews (R), letters, and early access (1975–2020). Note that the early-access document type covers articles that have been published electronically by a journal before being assigned to a specific volume and issue. Thus, early-access documents were amalgamated with articles or reviews. Proceedings papers published in a journal were considered as articles as well, while the other proceedings papers were considered as conference papers.

The results were double-checked to ensure that no irrelevant document was included. However, note that the search strings also excluded several relevant documents, which was usually because of the over-elimination of NOT keywords or typographical errors in the papers. For example, the string "lunar" was selected to exclude dozens of irrelevant documents, but consequently so was [53] related to AE; therefore, it had to be retrieved as part of the search results. Consequently, the search strings were cross-checked, leading to a further 59 relevant articles being retrieved.

All these actions yielded 9469 documents from 1975 to February 28, 2020. Table 1 gives their counts and corresponding percentage shares. Among these four types of paper, articles (73.23%) and conference papers (25.04%) were ranked first and second, respectively, in terms of the amount of papers. Text files of the full record and cited references of these documents were then downloaded to conduct the following bibliometric analysis.

Table 1. Detailed statistics on main document types regarding acoustic emission (AE) published in Web of Science (WoS) between 1975 and 2020-02-28.

Main Document types	Count	Share (%)
Article	6935	73．23
Review	82	0.87
Letter	81	0.86
Conference paper	2371	25.04
Total	9469	100

2.3. Basic principles of analytical tool and bibliometric methods
This work was carried out using the specialized bibliometric analysis tool VOSviewer, which was developed originally by van Eck and Waltman [54]. VOSviewer specializes in bibliometric network visualization by using advance scientometric techniques for layout and clustering, natural language processing, and creating bibliometric networks based on the data of the WoS or other global citation databases.

Natural language processing is used to create co-occurrence networks, which refer to multiple occurrences of wording or information described in document features such as the title, abstract, or author keywords. Author-keyword co-occurrence analysis is based on the frequency of pairwise author-given keywords in a data set with a large number of documents.

VOSviewer can also create co-authorship, bibliographic coupling, and co-citation networks to present bibliometric relationships among authors, documents, and cited references. Co-authorship is used mainly to analyze the co-signatures of authors in published papers. The logic of co-authorship is straightforward in that if two authors co-sign in one published paper, then they are considered to have a mutually cooperative relationship, which is expressed as a link (with a link strength of 1) connecting the two authors (nodes) in a co-authorship network. The bibliographic coupling of documents is used to measure the similarity of different documents by calculating the number of identical documents cited in them. Currently, the theory of bibliometric coupling analysis has been expanded from the original document coupling to author coupling, organization coupling, and country coupling to measure the research similarity or correlation between two authors, organizations, or countries. A co-citation relationship between two documents is established if they are simultaneously cited by a third document. The more frequently two documents are cited simultaneously, the closer is their academic relationship. The core principle of VOSviewer is VOS mapping and clustering. VOS mapping is based on constructing what are known as association normalization similarity matrices [55,56]. For a network of n nodes, $s_{ij} \in S$ denotes the similarity or association strength of nodes i and j and is calculated as:

$$s_{ij} = \frac{2mc_{ij}}{s_i s_j}, \forall i, j - 1,2, \ldots, n,$$

where m is the total number of links in the network, s_i is the total number of links of node i, and $c_{ij} \in C$ is the number of links (e.g., co-occurrence links, co-citation links, or bibliographic coupling links) between nodes i and j [55]. Once the similarity matrix is constructed, VOS mapping uses distance-based maps to find the location of node i in a p-dimensional map (usually p = 2) [55]. Meanwhile, VOS clustering uses a method that is similar to modularity-based clustering to conduct clustering analysis through maximizing [55]

$$V(c_1, \ldots, c_n) = \frac{1}{2m}\sum_{i<j} \delta(c_i, c_j) w_{ij} \left(c_{ij} - \gamma \frac{c_i c_j}{2m}\right),$$

where $w_{ij} = 2m/c_i c_j$, c_i is the cluster of node ith, the value of $\delta(c_i, c_j)$ is 1 if $c_i = c_j$ or 0 otherwise, and γ is the clustering resolution.

Based on the similarity or association strength between nodes i and j, Waltman et al. [55] proposed a unified approach of VOS mapping and clustering. Some important parameters that are critical to optimizing VOS mapping and clustering are defined. For example, the parameters of mapping attraction and repulsion are used to adjust and optimize the two-dimensional layout of the nodes in the map. The clustering resolution (γ) is associated with clusters of different resolutions, and a large γ will result in more subdivided clusters. These parameters will be identified specifically when a graphical visualization network is presented.

The principle of density mapping involves the methods of density view and clustering density view. In density view, the colour of a node in a map is determined according to the node density of that point, which depends on both the number and weights of its neighbouring nodes. For clustering density view, the node density of a point on a map is calculated separately for each cluster, and then the colour of a node on a map is determined.

3. Results

3.1. Publication trends

The amount of documents was accumulated every three years from 1975 to 2020. Fig. 1(a) shows the results of the overall trend of all document types as well as the two main types, namely article and review. The blue columns shaded with diagonal lines represent all document types (as listed in Table 1), while the red unshaded columns represent academic papers (article and review). Three conclusions can be drawn from Fig. 1(a). 1) Over the past 45 years, the overall trends of all document types and academic papers have been increasing year by year. 2) The middle of 2006 was the watershed for all AE-related papers: the amount of documents had been growing steadily before the middle of 2006, after which AE-related papers entered a stage of rapid development. 3) In the 1970s and 1980s, most of the document types were articles and reviews, after which the amount of other document types has risen over time. However, articles and reviews have always accounted for a large proportion of all document types: more than half of the documents are of these two types.

Using the co-occurrence analysis in VOSviewer, these papers were sorted into five types of document clusters with five different subjects. The subject names were determined by analyzing preliminary keywords in AE-related papers exported from WoS (see Section 3.2). These five subjects are (i) mechanical engineering, (ii) architectural and civil engineering, (iii) mining geophysics, (iv) metallurgical engineering, and (v) materials science and engineering. The detailed trend charts of these five different clusters are shown in Fig. 1(b)–(f). The overall trends of the five bar graphs are similar to those of Fig. 1(a): the number of papers in each cluster is increasing year by year; the middle of 2006 was the watershed for each cluster; articles and reviews account for the majority of the papers in each cluster.

Regarding the growing rate in the amount of papers, articles published in cluster 3 (mining geophysics) have the highest publishing rate, while the publishing rates of the other four themes are too close to call, indicating that the majority of scholars' attention has been on AE research for mining geophysics.

According to the number of published papers, the AE research in cluster 4 (metallurgical engineering) began earlier than did that on the other four subjects. Compared with the other four clusters, more AE-related papers in metallurgical engineering were published in the 1970s and 1980s. In comparison, fewer papers from cluster 4 were published in recent years than the other four subjects, which indicates that AE research on metallurgical engineering has matured. Meanwhile, papers published in cluster 3 (mining geophysics) have exceeded those in the other four subjects in recent years, showing that there are still many newly developing topics in this subject that are worthy of in-depth study by scholars.

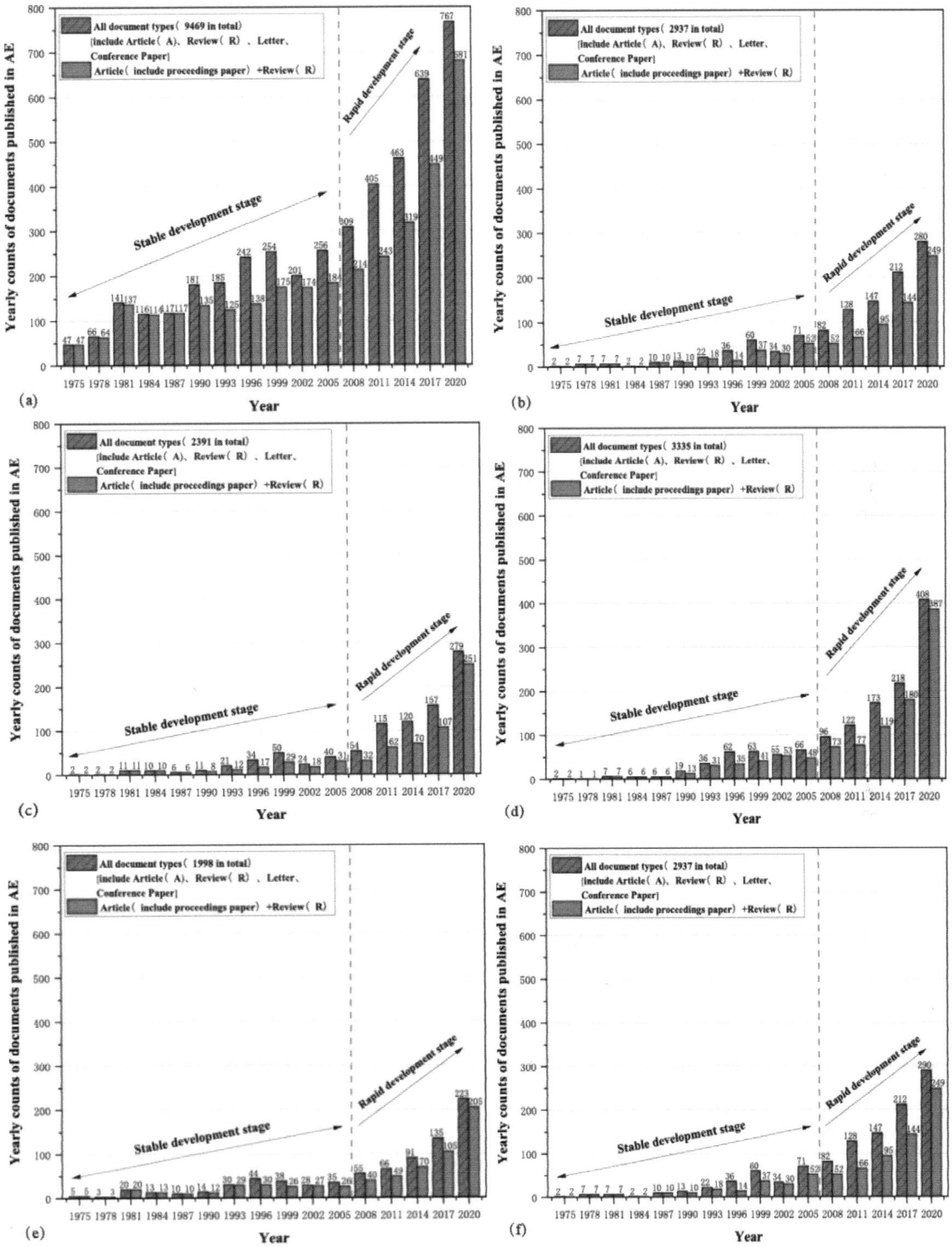

Figure 1. Annual quantities of all document types and academic papers (A+R): (a) overall trends; publication trends of (b) cluster 1 (mechanical engineering), (c) cluster 2 (architectural and civil engineering), (d) cluster 3 (mining geophysics), (e) cluster 4 (metallurgical engineering), and (f) cluster 5 (materials science and engineering).

3.2. Key areas of AE research

3.2.1. Clustering analysis of subjects. The purpose of the clustering analysis of subjects is to analyze and name various fields of AE over the past 45 years with the article classification method of the WoS and the co-occurrence method from VOSviewer. The latter is to analyze and classify the subjects of title and abstract using clustering algorithms. The reasons why we ignored the keywords WoS or Scopus has provided should be divided into two facets. On the one hand, although there are various statistical analyses of subject words in WOS or Scopus, many results from these platforms are directly shown towards learners without cautious selection. On the other hand, they provide crude results without much analysis of the results, which is exactly what we did in this article.

To exclude keywords with irrelevant clustering, we set the threshold value as 20 to filter out those keywords with fewer than 20 occurrences in the titles and abstracts of the AE-related papers. The rest of the parameters kept their default values. Finally, 657 keywords were generated in total. The VOS system divided these keywords automatically into five clusters, as shown in Fig. 2. The size of a circle represents the number of times that that keyword co-occurs, and the lines between keywords represent links. We screened out the top 20 keywords in each cluster and analyzed the research subjects of the five clusters from the papers in the WoS. Finally, we found that the five clusters could be termed mechanical engineering (red, cluster 1), architectural and civil engineering (green, cluster 2), mining geophysics (blue, cluster 3), metallurgical engineering (yellow, cluster 4), and materials science and engineering (purple, cluster 5). The representative keywords of each cluster are listed below, but these are only for data cluster analysis and do not contain theoretical knowledge.

Fig. 2 shows that cluster 1 is the largest, indicating that it has the largest quantity of co-occurring keywords. The topics in this cluster are broader than those in the other clusters. In other words, the topics in clusters 2-5 are concentrative. In addition, Fig. 2 shows that the circles of clusters 2–4 are interlaced with each other, which reveals the relatedness of these subjects.

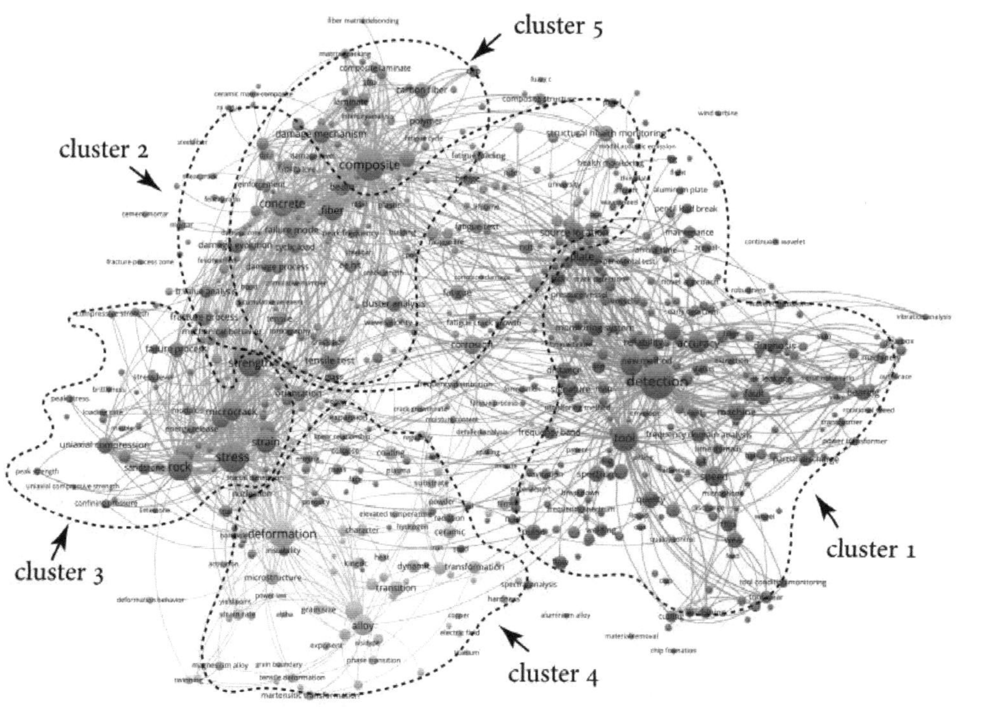

Figure 2. Network visualization of key AE research areas from 1975 to 2020.

Table 2. Top five co-occurring keywords from mechanical engineering.

No.	Keyword	Times of occurrence
1	detection	999
2	tool	577
3	defect	459
4	accuracy	334
5	plate	319

The keywords in cluster 1 (mechanical engineering) include detection, defect, tool, machine, diagnosis, accuracy, and fault. The sizes of these circles show that the keywords that co-occur most frequently include detection, tool, and defect. As listed in Table 2, the top five co-occurring keywords from cluster 1 represent some common research interests of AE in fault diagnosis, condition monitoring, source location, and process monitoring in mechanical engineering.

The co-occurring keywords in cluster 2 (architectural and civil engineering) include fracture, loading, rock, concrete, and stress. As listed in Table 3, the five most frequently used keywords relate mainly to the testing, evaluating, detecting, and monitoring of concrete materials and structural health.

The co-occurring keywords in cluster 3 (mining geophysics) include rock, earthquake, and microcrack. As listed in Table 4, the top five terms from this subject relate to the application of AE to rock testing and seismic monitoring.

The co-occurring keywords in cluster 4 (metallurgical engineering) include alloy, deformation, corrosion, stainless steel, microstructure, and dislocation. The top five terms from this subject are listed in Table 5.

Table 3. Top five co-occurring keywords from architectural and civil engineering.

No.	Keyword	Times of occurrence
1	Concrete	530
2	Corrosion	273
3	fracture process	231
4	beam	229
5	fatigue	199

Table 4. Top five co-occurring keywords from mining geophysics.

No.	Keyword	Times of occurrence
1	stress	743
2	Rock	558
3	Strength	545
4	Strain	393
5	Microcrack	341

Table 5. Top five co-occurring keywords from metallurgical engineering.

No.	Keyword	Times of occurrence
1	deformation	596
2	alloy	343
3	plastic deformation	287
4	transition	182
5	microstructure	172

The co-occurring key words in cluster 5 (materials science and engineering) include composite, fiber, damage mechanism, and tensile test. As listed in Table 6, the keyword that co-occurs most frequently is composite. This cluster is related to the testing, evaluating, and monitoring of fibers, glass-fiber reinforced plastic, and carbon-fiber reinforced plastic (CFRP), as well as other composite materials. In materials science and engineering, AE is known as a powerful NDT tool.

3.2.2. Thematic impact analysis of AE topics. The purpose of the thematic impact analysis is to reveal the influence and activity of AE topics and different subjects. This can be realized through evaluating the average publishing year and the average number of citations for each AE topic based on co-occurring keywords using VOSviewer. As shown in Fig. 3, the size of a circle represents the number of times that that word co-occurs. The color of a circle shows the average publishing year of that keyword: the closer the color is to yellow, the more recent the attention paid to that topic; the closer the color is to blue, the further back in time that topic was researched.

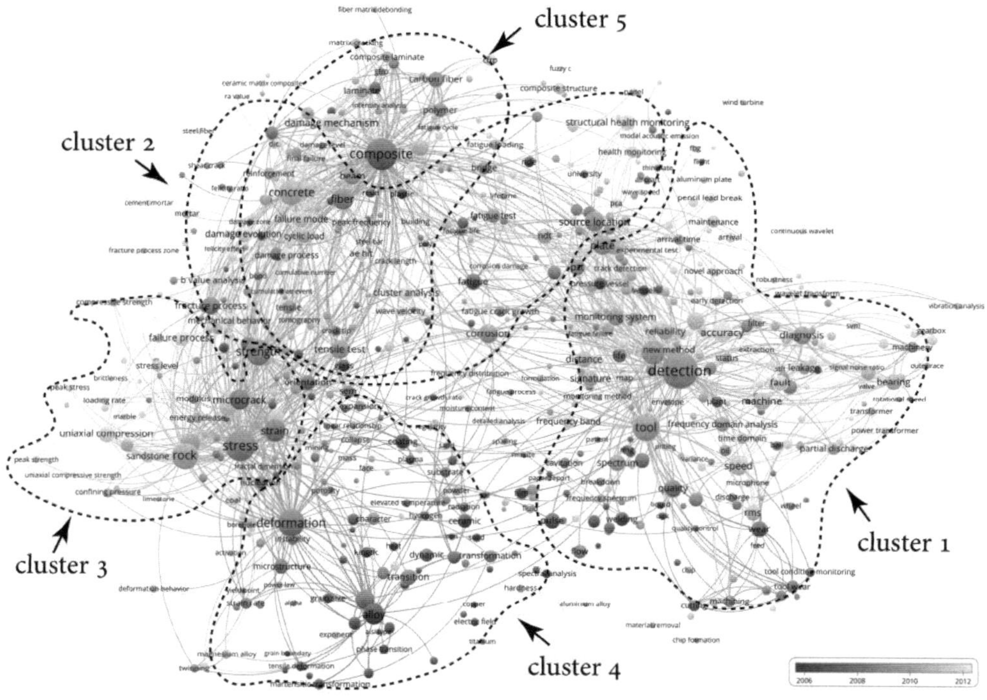

Figure 3. Overlay visualization of average publishing year of the five clusters.

Fig. 3 shows that the colors of cluster 4 (metallurgical engineering) are darkest among the clusters, showing that articles from this subject have been few in number in recent years. This reflects the fact that research on this topic is relatively mature. For example, the phase-change detection of steels, such as martensitic transformation detection and dislocation using AE, is now well established.

Regarding the other four subjects, some newly developing topics are attracting increasing numbers of current scholars. In cluster 1 (mechanical engineering), topics such as vessel, welding, and detection are less active. The most commonly used high-frequency keywords still occupy a great proportion of all keywords, such as accuracy, gearbox, fault diagnosis, online monitoring, and arrival, indicating that the most active AE research in mechanical engineering in recent years has been focused on these topics.

In cluster 2 (architectural and civil engineering), topics concerning SHM, dic (digital image correlation), damage process, mechanical behaviour, and AE hit have been popular in recent years. By contrast, the topic of fatigue was active before 2008.

Table 6. Top five co-occurring keywords from materials science and engineering.

No.	Keyword	Times of occurrence
1	composite	790
2	fiber	356
3	damage mechanism	290
4	tensile test	283
5	delamination	233

The overall color of cluster 3 (mining geophysics) is closer to yellow, indicating that research on this subject has been active in recent years. Fig. 3 shows that AE research on failure process, uniaxial compression, mining, and coal has been active since 2012.

In cluster 5 (materials science and engineering), the topics of damage monitoring, damage mechanism, CFRP, fuzzy clustering, cluster analysis, and damage evolution have been active recently, whereas AE studies on plastic and glass have been less active.

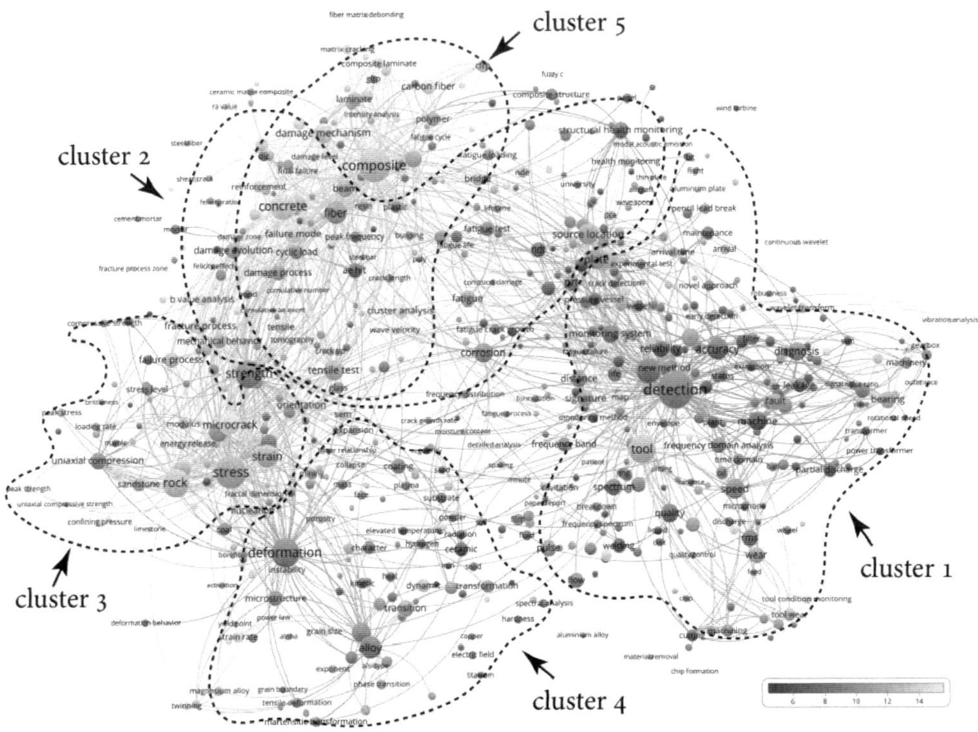

Figure 4. Occurrence overlay visualization of normalized average number of citations of the five clusters.

Fig. 4 presents the overlay visualization of the average number of citations of these AE topics. The size of a circle represents the total number of citations, while the color represents the average number of citations. Clearly, the average numbers of citations of cluster 2 (architectural and civil engineering), cluster 3 (mining geophysics), and cluster 5 (materials science and engineering) are significantly higher than those of the other two clusters, indicating that AE has a high degree of research activity in these subjects. By contrast, the average numbers of citations of keywords in clusters 1 and 4 are significantly lower than those of the other subjects. Nevertheless, AE topics such as tool condition monitoring, gearbox, and chip formation in cluster 1, as well as twining and grain size in cluster 4, are highly cited, indicating the active study of AE in these areas.

IOP Conf. Series: Materials Science and Engineering 1167 (2021) 012009 doi:10.1088/1757-899X/1167/1/012009

3.3. Organization analysis

3.3.1. Leading organizations. The 10 organizations that publish the most AE articles are listed in Table 7, most of which are in Japan, China, and the United States of America (USA). The top three organizations are the University of Tokyo, the Russian Academy of Sciences, and China University of Mining and Technology.

Table 7. Organizations contributing the most AE papers.

No.	Organization	Country/Region	TP	TC	TC/TP
1	Univ Tokyo	Japan	134	1044	7.79
2	Russian Acad Sci	Russia	124	591	4.77
3	China Univ Min and Technol	China	121	1319	10.90
4	Cranfield Univ	England	97	1580	16.29
5	Chinese Acad Sci	China	84	560	6.67
6	Charles Univ Prague	Czech Republic	68	945	13.90
7	Kumamoto univ	Japan	68	1579	23.22
8	Northeastern univ	USA	66	323	4.89
9	Univ Illinois	USA	60	747	12.45
10	Kyoto Univ	Japan	58	828	14.28

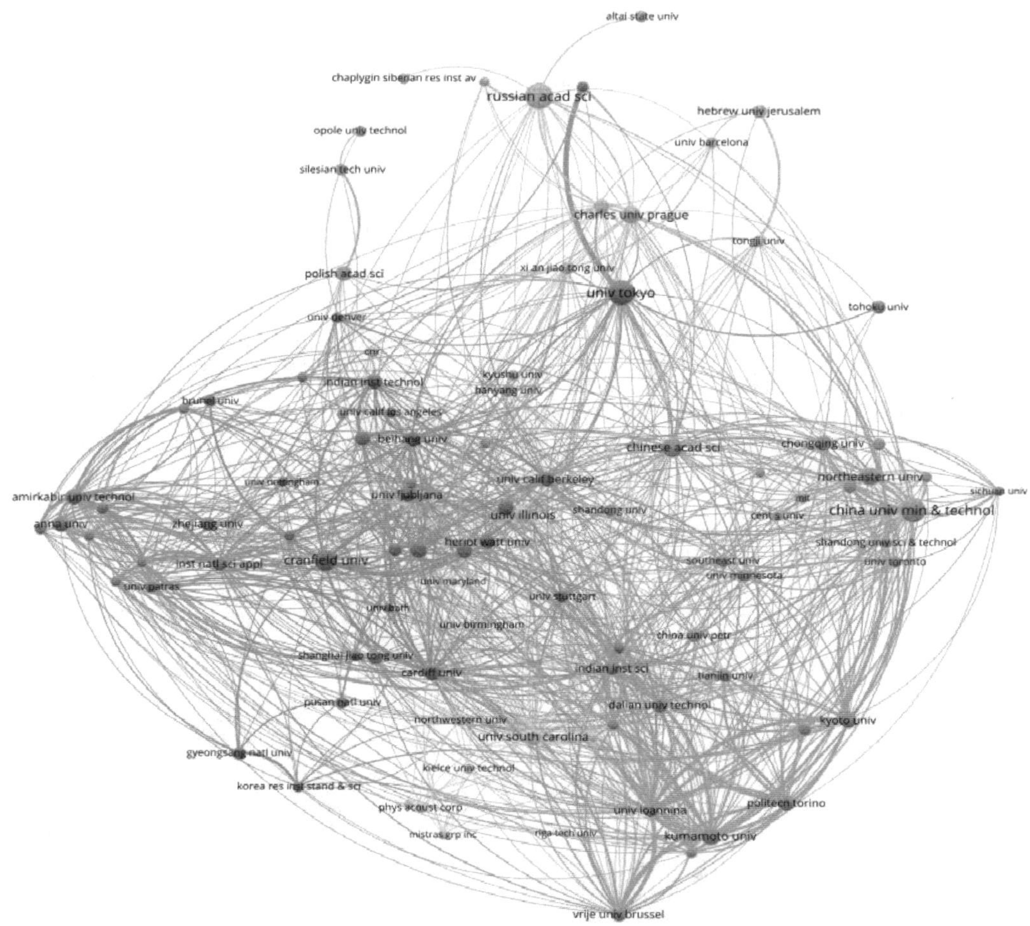

Figure 5. Citation network of various organizations.

3.3.2. Citation analysis of organizations. The threshold of articles published by each organization is set to 20, giving 100 organizations in total. The results are presented in Fig. 5, where the size of a circle represents the number of articles published by that organization. Organizations with strong citation links are clustered together, giving eight interlaced clusters. This shows that AE promotes mutual exchange and cooperation among different organizations. The most-cited organizations from clusters 1–8 are Charles University Prague, University of Tokyo, Polish Academy of Science, Institution of National Science and Application, Cranfield University, Kumamoto University, University of South Carolina, and China University of Mining and Technology, respectively. In the purple cluster, apart from Kumamoto University, there is another distinguished Italian university, called Polytechnic University of Turin (Politecnico di Torino), which also has published a large amount of AE papers and in recent years has reached levels of publications on AEs comparable to the starting universities in the world.

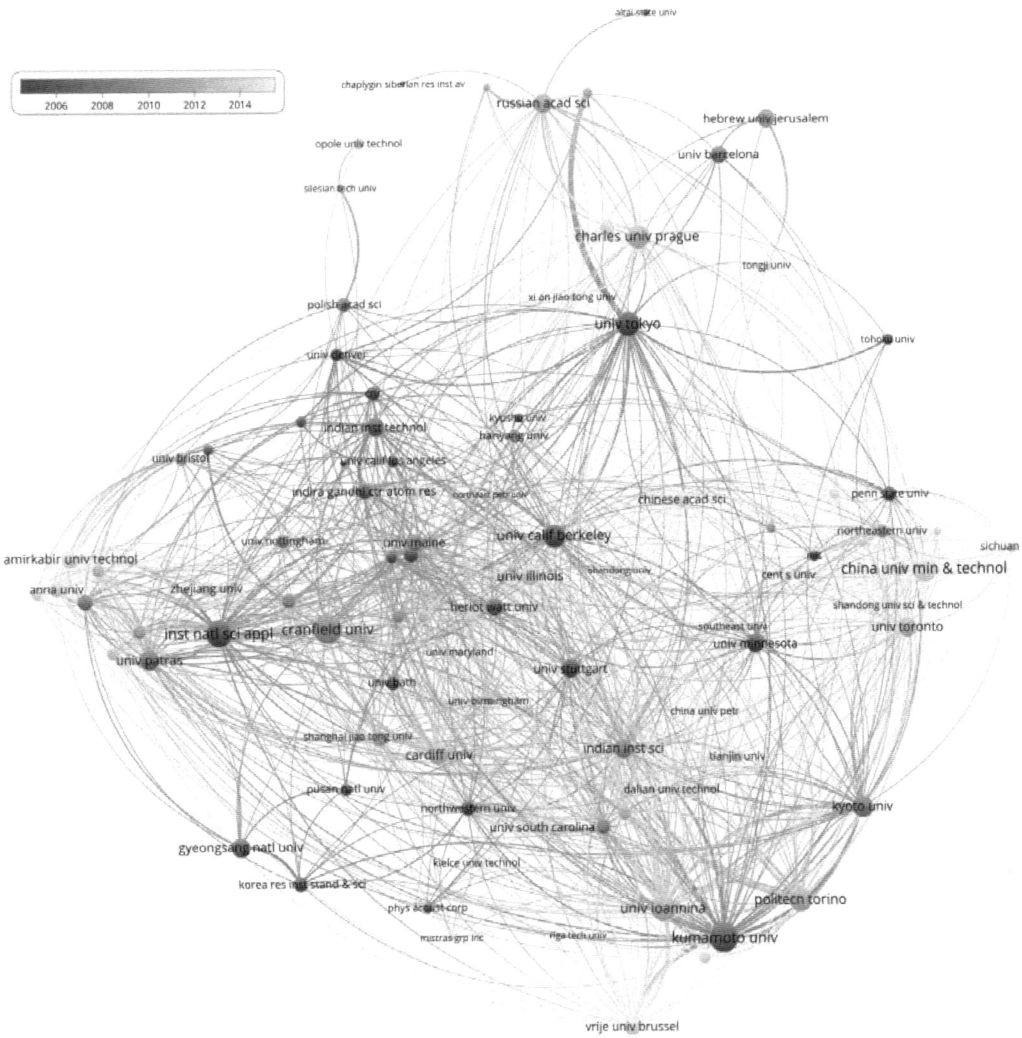

Figure 6. Overlay visualization of average publishing year of various organizations.

The overlay visualization of average publishing year in Fig. 6 shows that the time of AE research varies from organization to organization. University of Tokyo, University of California Berkeley, Inst Natl Sci App, Kumamoto University, University of Minnesota, and other organizations with dark blue circles carried out their AE studies before 2008, with few further AE studies taking place in those organizations thereafter. The green cluster, representing some organizations which started to research acoustic emission from 2010 to 2014, includes a lot of well-known universities, like University of Ioannina, Polytechnic University of Turin, Charles University, Russian academy of science and so on. In comparison, China University of Mining and Technology, Chinese Academy of Sciences, Amirkabir University of Technology, Vrije Universiteit Brussel, University of South Carolina, and other organizations with yellow circles have remained active in recent years. Indeed, some of those organizations only began to study AE in recent years.

Fig. 7 shows the visualization of the average number of citations of each organization. The size of a circle represents the total number of citations, and the color represents the average number of citations. A line between two circles represents a citation relationship, with the line width representing the citation intensity.

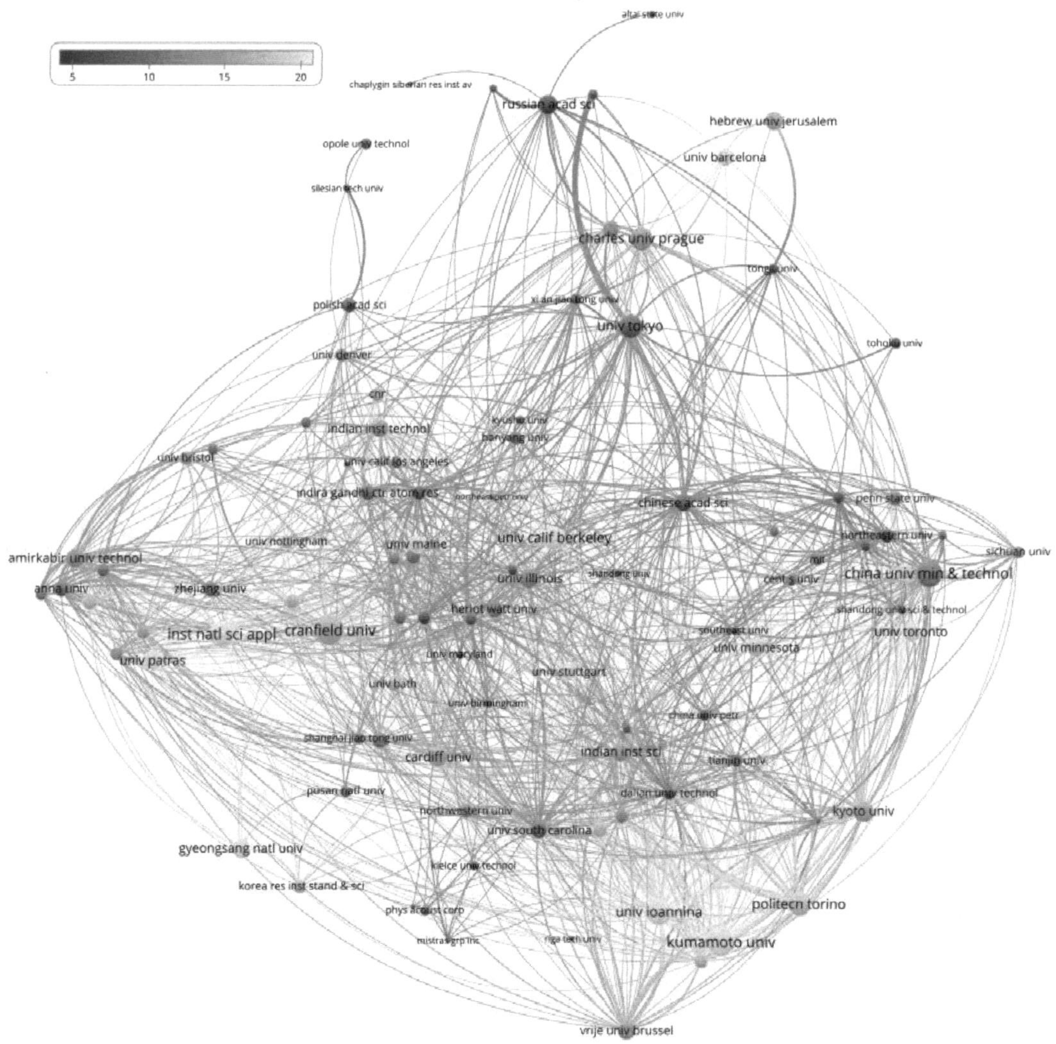

Figure 7. Overlay visualization of average number of citations of various organizations.

3.4. Analysis of countries

3.4.1. Leading countries. The 10 countries that publish the most AE papers are listed in Table 8, along with their total number of publications(TP), total number of citations(TC), TC/TP and h-index. From

Table 7, the organizations with the most publications are in China, USA, and Japan, and these countries also happen to be the top three countries in terms of published papers, with totals of 1604, 1247, and 765, respectively. However, the amount of citations among all of the papers these countries (TC/TP) have published rank differently compared with total publications(TP). The countries with relative big proportions of the amount of citations versus total publications are France, Germany, England and Italy.

Table 8. Countries contributing the most AE papers.

No.	Region/Country	TP	TC	TC/TP	h-index
1	China	1604	8720	5.44	43
2	USA	1247	15785	12.6	90
3	Japan	765	7842	10.25	50
4	England	590	8917	15.11	61
5	Russia	561	2251	4.01	28
6	Germany	350	5443	15.55	56
7	France	340	6355	18.69	56
8	India	337	3290	9.76	30
9	Italy	249	3563	14.31	46
10	Canada	196	2679	13.67	44

3.4.2. Citation analysis of countries. The threshold of articles published by each country is set as 10, giving 54 countries in total. Their citation network is shown in Fig. 8, where the size of a circle represents the number of published articles. A line between two circles represents a citation relationship, with the line width representing the citation intensity.

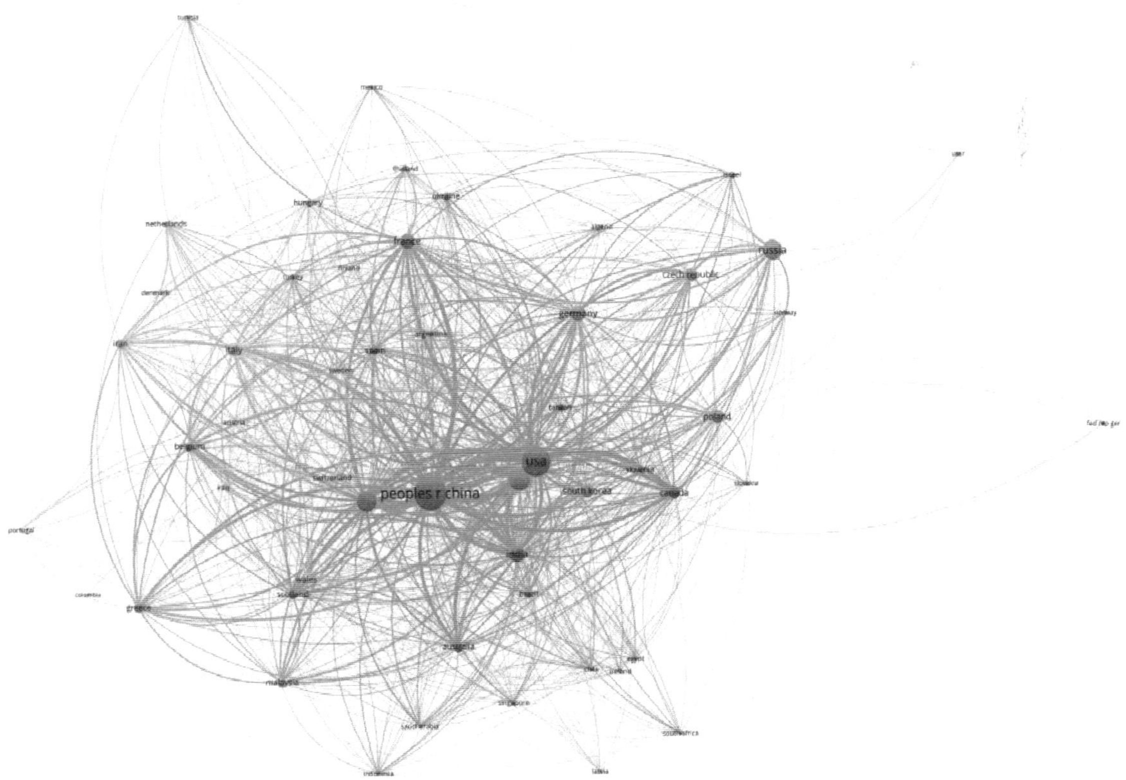

Figure 8. Citation network of various countries.

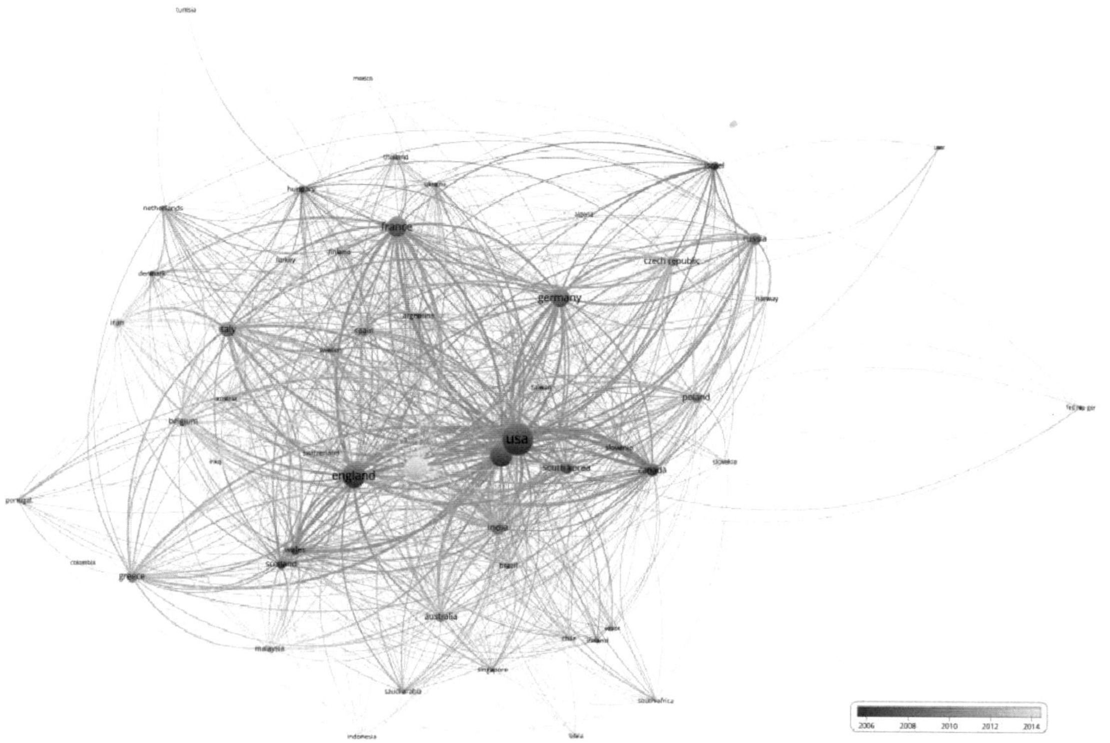

Figure 9. Overlay visualization of average publication year of various countries.

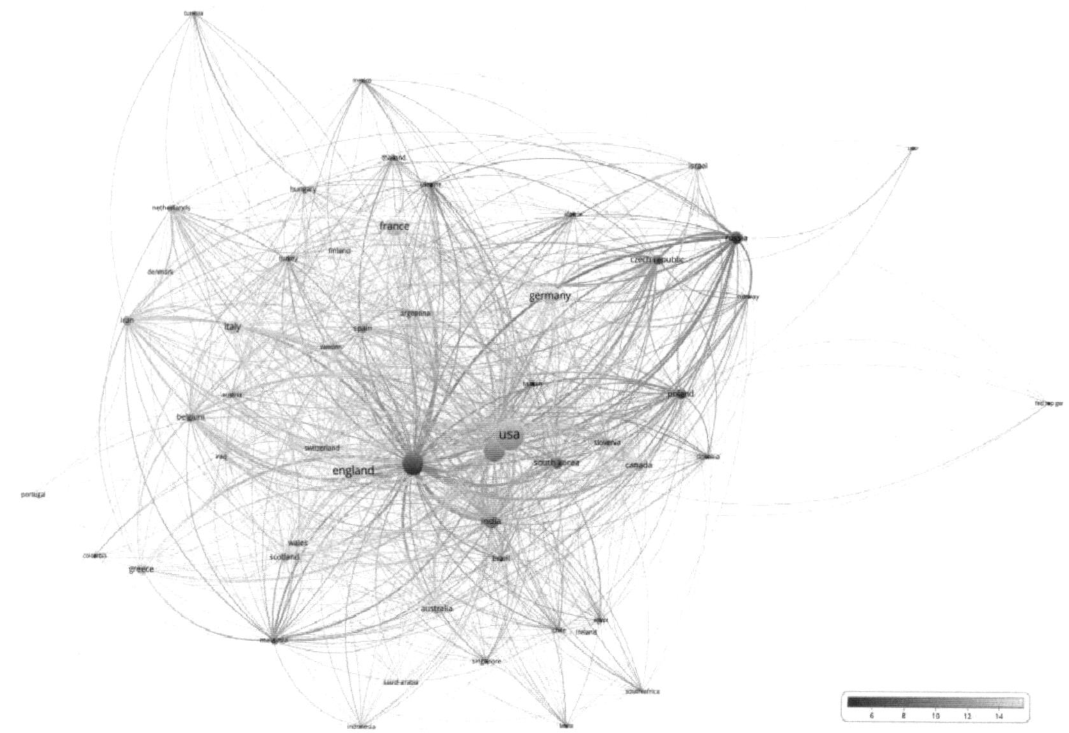

Figure 10. Overlay visualization of average number of citations of various countries.

Fig. 8 shows that USA, China, England, Japan, and Switzerland are located at the center of this network and are linked with almost every other country, indicating that articles published by these countries are cited widely. Note also that the red cluster contains 17 items, which is the largest among all five clusters. In comparison, the other four clusters share a similar proportion of approximately seven items.

Fig. 9 shows the average publication year of countries. Similarly, the size of a circle reflects the total number of citations, while the color reflects the average publication year of that country.

Clearly, the colors of most circles are between blue and green, indicating that the average publication year of most countries was around 2005. The average citation years of USA, Japan, England, Canada, and Israel are before 2008, while those of China, Iran, Iraq, Turkey, Mexico, Tunisia, Algeria, Malaysia, Columbia, and Indonesia are after 2014. Indeed, some are just beginning to conduct AE research. It can be concluded that research interest in AE is shifting to developing countries, caused by the rapid development of industry, engineering, and academic research in those countries. Fig. 10 shows the average number of citations of countries. Similarly, the size of a circle represents the total number of citations, while the color represents the average number of citations. The closer a circle is to yellow, the more citations that country receives each year.

It can be seen that although China and Russia have published many AE-related articles, they have the fewest cited papers per year. By contrast, Germany, France, Italy, Finland, Denmark, Portugal, England, Wales, Greece, Saudi Arabia, and Ireland received more citations per year, exceeding approximately 14. This indicates that the level of research in those countries is higher in terms of both quality and originality.

3.5. Citation analysis of sources

The reasons for analyzing the source journals of the AE literature are to track the latest scientific research information, to select the best-written journals, and to establish which journals have published the most items. The threshold number of papers published by each source was set as 20, giving 75 publication sources in total. The network visualization results are shown in Fig. 11, where the size of a circle represents the number of papers published by that source.

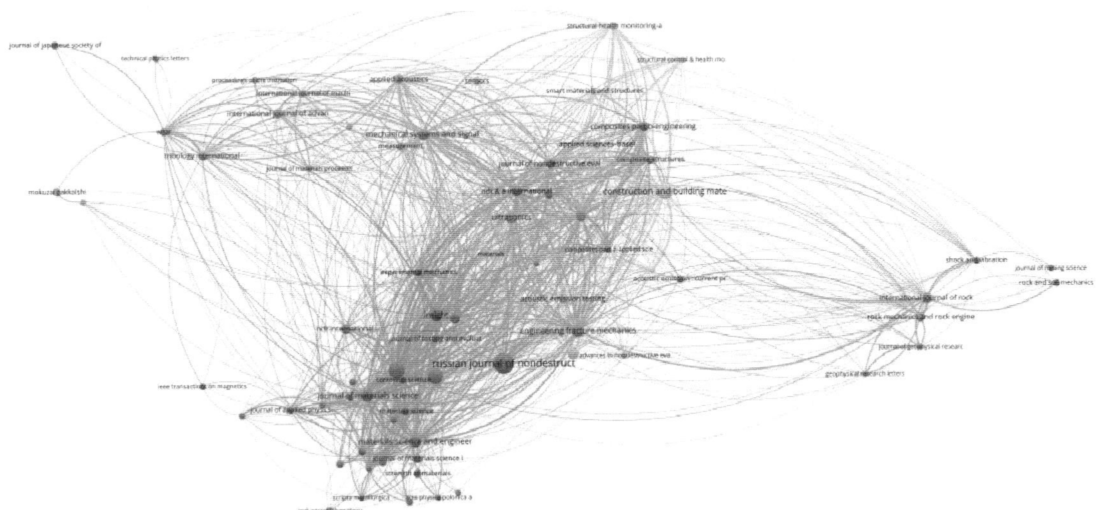

Figure 11. Citation network of various sources.

It can be concluded from Fig. 11 that each of the seven clusters belongs to a specific subject. For example, sources in the red cluster are focused on materials science and engineering, while those of the purple cluster have geophysical subjects. The red cluster contains the largest number of publications, among which the Russian Journal of Nondestructive and Materials Evaluation was the journal that published the most articles. The top five sources of AE publications are listed in Table 9.

Table 9 shows that sources with large numbers of documents are not necessarily those that are cited the most. The top five cited sources are listed in Table 10. We noted that Journal of Acoustic Emission (JAE) has published a large number of papers related to AE. However, since JAE is not included in the core journal of WOS, this journal and its published articles were not counted in this paper.

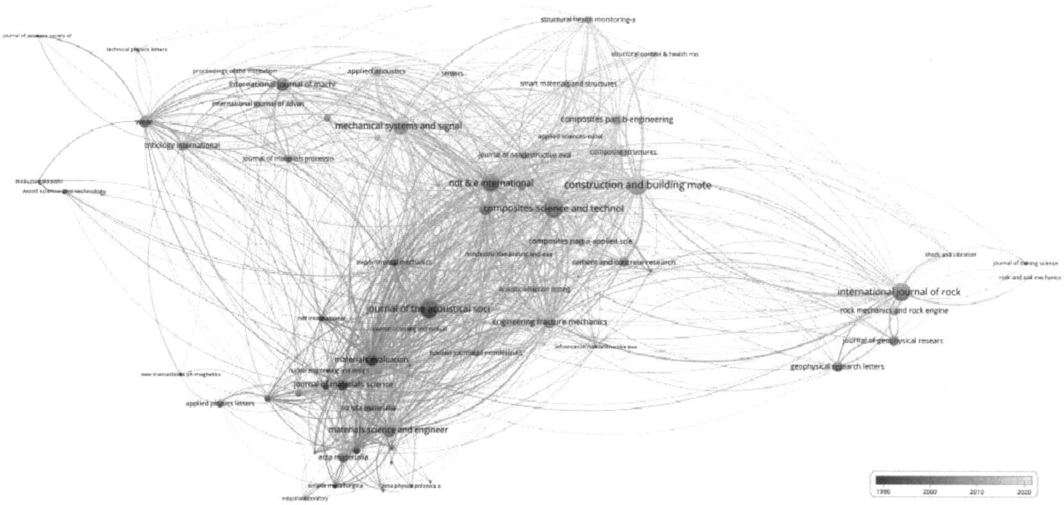

Figure 12. Overlay visualization of average publication year of various sources.

Fig. 12 shows a visualization of the average publication year of these sources. The size of a circle represents the total number of citations, while the color represents the average publication year of these sources. Clearly, most circles are light blue or bright yellow, indicating an average publication year after 2010. The average publication years of certain sources, e.g., Structural Health Monitoring, Applied Acoustics, Measurements, Sensors, Shock and Vibration, as well as Rock Mechanics and Rock Engineering, are even closer to the present. In comparison, some journals have an average publication year that is before 1990, such as Soviet Journal of Nondestructive Testing-USSR (renamed in 1991 as Russian Journal of Nondestructive Testing), Materials Evaluation, Materials Science and Engineering, NDT&E International, Industrial Laboratory, and Journal of the Acoustical Society of America.

Table 9. Top five publication sources.

R	Source	Documents	Citations
1	Russian Journal of Nondestructive and Materials Evaluation	238	408
2	Materials Evaluation	153	1204
3	Construction and Building Materials	140	3366
4	Soviet Journal of Nondestructive Testing-USSR	140	19
5	Engineering Fracture Mechanics	101	1265

Table 10. Top five cited sources.

R	Source	Documents	Citations
1	Construction and Building Materials	3366	140
2	Composites Science and Technology	3165	67
3	Journal of the Acoustical Society of America	2972	64
4	International Journal of Rock Mechanics and Mining Sciences	25994	64
5	NDT&E International	2416	70

Fig. 13 shows the overlay visualization of the average number of citations of each source. The size of a circle reflects the total number of citations, while the color reflects the average number of citations. This shows that Mechanical Systems and Signal Processing, International Journal of Machine Tools & Manufacture, NDT&E International, Composite Science and Technology, Composite Part A – Applied Science, Cement and Concrete Research, International Journal of Rock, Journal of Geophysical Research, Geophysical Research Letters, Journal of the Acoustical Society of America, Corrosion Science, and Acta Materialia each have over 30 citations on average. This indicates that articles published in these sources are higher in quality and originality than are those in other sources.

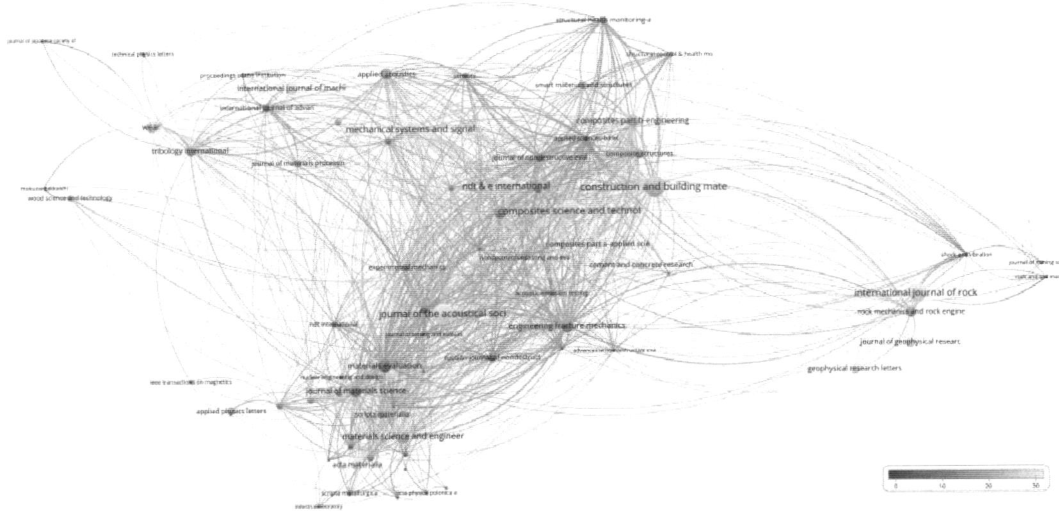

Figure 13. Overlay visualization of average number of citations of various sources.

4. Conclusions

The aim of this work was to conduct a comprehensive overview of the application of AE in various areas. Therefore, we carried out a bibliometric review based on the WoS database and the specialized bibliometric analysis tool VOSviewer. We analyzed the publication trends and key areas, organizations, countries, and publication sources of AE applications. The influences of subjects and AE topics were further discussed to provide a clear view of the current state of AE research. The influences of organizations, countries, and publication sources were also discussed. The main findings of the present work are summarized as follows.

1. The number of papers on AE applications has been rising rapidly since 2006, indicating that the application of AE remains a popular research issue.

2. The application of AE in mechanical engineering accounts for the largest proportion of published papers, whereas research on AE applications in metallurgical engineering is lacking space for further development.

3. Active in recent years have been the AE topics of (i) fault diagnosis, condition monitoring, and quality control in mechanical engineering, (ii) SHM, damage processes, and mechanical behavior in architectural and civil engineering, (iii) failure processes, uniaxial compression, and mining in mining geophysics, and (iv) damage monitoring, damage mechanisms, and clustering analysis in materials science and engineering.

4. The leading organizations, countries, and publication sources were provided, and their current states and potential influences were discussed.

In this paper, we used VOSviewer, a convenient visualization software, to systematically analyze the research status and development trend of AE, and the retrieval method of this paper would make a more careful and comprehensive reference for the subsequent scientific research scholars on literature information retrieval. Faced with the topics or academic fields of interest, people can use the software to conduct visual analysis on the information of different topics and academic fields, so as to quickly achieve the hot trends and key fields in recent years. It can also compare the influence of different publishers or organizations, which greatly saves the time of literature retrieval for readers and researchers, and also improves the efficiency of scientific research.

Acknowledgments
This project was supported by the National Natural Science Foundation of China (Grant No. 51905335) and the Shanghai Science and Technology Committee Research Project (Grant No. 19040501500).

References

[1] ASTM E1316-20 2020 Standard Terminology for Nondestructive Examinations *ASTM Inte.* **PA**
[2] Kishawy H A, Hegab H, Umer U and Mohany A 2018 Application of acoustic emissions in machining processes: analysis and critical review *Int. J. Adv. Manuf. Tech.* **98** 1391-1407
[3] Lee DE, Hwang I, Valente CMO, Oliveira JFG, and Dornfeld DA 2006 Precision manufacturing process monitoring with acoustic emission *Int. J. Mach. Tool. Manu.* **46** 176-188
[4] Xiaoli Li 2002 A brief review: acoustic emission method for tool wear monitoring during turning *Int. J. Mach. Tool. and Manu.* **42** 157-165
[5] Kishawy HA, Hegab H, Umer U, and Mohany A 2018 Application of acoustic emissions in machining processes: analysis and critical review *Int. J. Adv. Manu. Tech.* **98** 1391-1407
[6] Jayakumar T, Mukhopadhyay CK, Venugopal S, Mannan SL, and Baldev Raj 2005 A review of the application of acoustic emission techniques for monitoring forming and grinding processes *J. mater. process. tech.* **159** 48–61
[7] Kanji O 2018 Review on structural health evaluation with acoustic emission *Appl. Sci.* **8** 958
[8] Kanji O 2005 Current understanding of mechanisms of acoustic emission *J. Strain Anal. Eng.* **40** 1-15
[9] Kanji O 2011 Application of acoustic emission for structure diagnosis *Diagnostyka* 3-18
[10] Kanji O and Gallego A 2012 Research and applications of ae on advanced composites *J. Acous. Emi.* **30** 180-229
[11] Shiotani T 2012 Recent advances of ae technology for damage assessment of infrastructures *J. Acous. Emi.* **30**
[12] Robert HD, Cavaco JA and Mustafa V 2009 Monitoring the civil infrastructure with acoustic emission: bridge case studies *J. Acous. Emi.* **27**
[13] Manthei G and Plenkers K 2018 Review on in situ acoustic emission monitoring in the context of structural health monitoring in mines *Appl. Sci.* **8** 1595
[14] Sengupta S, Kumar DA, and Topdar P 2015 Structural damage localisation by acoustic emission technique: A state of the art review *Lat. Am. J. Solids. Stru.* **12** 1565-82
[15] Holford KM, Mark JE, Hensman JJ, Pullin R, Evans SL, Dervilis N, and Worden K 2017 A new methodology for automating acoustic emission detection of metallic fatigue fractures in highly demanding aerospace environments: an overview *Prog. Aerosp. Sci.* **90** 1-11
[16] Nair A and Cai CS 2010 Acoustic emission monitoring of bridges: review and case studies *Eng. stru.* **32** 1704-14
[17] A. Carpinteri, G. Lacidogna, N. Pugno 2007 Structural damage diagnosis and life-time assessment by acoustic emission monitoring *Eng. Frac. Mech.* **74** 273-289
[18] Sagar RV and Prasad BKR 2012 A review of recent developments in parametric based acoustic emission techniques applied to concrete structures *Nondestruct. Test. Eva.* **27** 47-68
[19] Carpinteri A, Lacidogna G, and Pugno N 2007 Structural damage diagnosis and life-time assessment by acoustic emission monitoring *Eng. Frac. Mech.* **74** 273-289
[20] Kurz JH, Grosse CU, and Reinhardt HW 2005 Strategies for reliable automatic onset time picking of acoustic emissions and of ultrasound signals in concrete *Ultrasonics* **43** 538-546
[21] Noorsuhada MN 2016 An overview on fatigue damage assessment of reinforced concrete

structures with the aid of acoustic emission technique *Constr. Build. Mater.* **112** 424-439
[22] Tsangouri E and Aggelis DG 2019 A review of acoustic emission as indicator of reinforcement effectiveness in concrete and cementitious composites *Constr. Build. Mater.* **224** 198–205
[23] Ahmed AN, Soliman AA, Khider SE, Casey NF and Laura PAA 1997 A review of the acoustic emission monitoring of wire rope *Ocean Eng.* **24** 935-947
[24] Drummonda G, Watsonb JF, Acarnleyb PP 2007 Acoustic emission from wire ropes during proof load and fatigue testing *NDT Eng. Int.* **40** 94-101
[25] Rosa IMD, Santulli C, and Sarasini F 2009 Acoustic emission for monitoring the mechanical behaviour of natural fibre composites: a literature review *Appl. Sci. Manu.* **40** 0-1469
[26] Behnia B, Buttlar W, and Reis H 2018 Evaluation of low-temperature cracking performance of asphalt pavements using acoustic emission: a review *Appl. Sci.* **8** 306
[27] Romhány G, Czigány T, and Kocsis JK 2017 Failure assessment and evaluation of damage development and crack growth in polymer composites via localization of acoustic emission events: a review *Polym. Rev.* **57** 397-439
[28] Faisal NH, Ahmed R, and Reuben RL 2011 Indentation testing and its acoustic emission response: applications and emerging trends *Int. Mater. Rev.* **56** 98-142
[29] Michlmayr G, Cohen D, and Dani O 2012 Sources and characteristics of acoustic emissions from mechanically stressed geologic granular media:a review *Earth-Sci. Rev.* **112** 97-114
[30] Nejati HR, Nazerigivi A, Mehrdad I, and Karrech A 2020 Monitoring of fracture propagation in brittle materials using acoustic emission techniques-a review *Compu. Concrete* **25** 15
[31] Soberanis CR 2011 Acoustic emission technique, an overview as a characterization tool in materials science *J. appl. Re. Tech.* **9** 367-379
[32] Salah M, Obaidi A, Leong MS, Hamzah RIR, and Abdelrhman AM 2012 A review of acoustic emission technique for machinery condition monitoring: defects detection and diagnostic *Appl. Mech. Mater.* **229** 1476-80
[33] Caesarendra W, Kosasih B, Tieu AK, Hongtao Z, and Zhu Q 2015 Acoustic emission-based condition monitoring methods: review and application for low speed slew bearing *Mech. Sys. Sig. Process.* **72** 134-159
[34] Mylvaganam S 2003 Some applications of acoustic emission in particle science and technology *Parti. sci. tech.* **21** 293-301
[35] Tan CK and David Mba 2005 Limitation of acoustic emission for identifying seeded defects in gearboxes *J. NDT. Eva.* **24** 11-28
[36] Tim T, Tan CK, and David M 2005 Application of acoustic emission to seeded gear fault detection *Ndt. Eng. Int.* **38** 27-36
[37] Yan J et al. 2015 Nondestructive detection of valves using acoustic emission technique *Adv. Mater. Sci. Eng.*
[38] Tinghu Y and Jones BE 2004 Acoustic emission testing on large structures *Meas. Control* **37** 168-173
[39] Jonathan WRB and Varley J 2001 The uses of passive measurement of acoustic emissions from chemical engineering processes *Chem. Eng. Sci.* **56** 1749-67
[40] Gholizadeh S, Leman Z, and Baharudin BTHT 2015 A review of the application of acoustic emission technique in engineering *Stru. Eng. Mech.* **54** 1075-95
[41] Krzysztof J 2001 Some aspects of acoustic emission signal pre-processing *J. Mater. Process. Tech.* **109** 242-7
[42] Terchi A and Au YHJ 2001 Acoustic emission signal processing *Meas. Control* **34** 240-244
[43] Liang Z, Le K and Shuang Y 2018 Research and application of acoustic emission signal processing technology *IEEE Access* **7** 984-93
[44] Richard AK 2016 Acoustic emission in orthopaedics: a state of the art review *J. biomech.* **49** 4065-72
[45] Mustafa SR and Rhys Pn2014 The sound of orthopaedic surgery—the application of acoustic emission technology in orthopaedic surgery: a review *Euro. J. Ortho. Surg. Trauma.* **24** 1-6

[46] Browne M, Roques A and Taylor A 2005 The acoustic emission technique in orthopaedics-a review *J. Strain Anal. Eng. Design* **40** 59-79
[47] Olorunlambe KA, Shepherd DET and Dearn KD 2019 A review of acoustic emission as a biotribological diagnostic tool *Tribo. Mater. Surf. Interfaces* **13** 161-71.
[48] Fereidoon MN, Solmaz K and Zakeri H 2019 A brief review on acoustic analysis in quality evaluation and a new method for determining bulk density of aggregate *Arch. Compu. Method. Eng.* **26** 1577-91
[49] Tanjin AMd, Faisal K and Paul A 2019 A bibliometric review of process safety and risk analysis *Process Saf. Environ.*
[50] Xin Z, Wen LY and Hai LV 2018 Visualization and analysis of mapping knowledge domain of road safety studies *Accident Anal. Prev.* **118** 131-45
[51] Matthew EF, Eleni IP, George AM and Georgios P 2008 Comparison of pubmed, scopus, web of science, and google scholar: strengths and weaknesses *FASEB J* **22** 338-42
[52] Johannes V, Stefan PE and Doncker RWD 2019 Control method for avoiding transformer saturation in high-power three-phase dual-active bridge dc–dc converters *IEEE Trans. Power Electro.* **35** 4332-41
[53] Quan L et al. 2019 Correlations between acoustic and electromagnetic emissions and stress drop induced by burst-prone coal and rock fracture *Saf. Sci.* **115** 310-19
[54] Nees VE and Ludo W 2010 Software survey: Vosviewer, a computer program for bibliometric mapping *Scientometrics* **84** 523-38
[55] Ludo W, Eck NJV and Noyons ECM 2010 A unified approach to mapping and clustering of bibliometric networks *J. Info.* **4** 629-35
[56] Nees JVE and Ludo W 2009 How to normalize cooccurrence data? an analysis of some well-known similarity measures *J. Ame. Soci. Info. Sci. Tech.* **60** 1635-51

IOP Conf. Series: Materials Science and Engineering 1167 (2021) 012010 doi:10.1088/1757-899X/1167/1/012010

Temperature dependent performance of lead phthalocyanine phototransistors

Yao Li[1,2], Miao Pan[1], Zheng Zhang[1], Rong Wu[1], Caizhen Zhang[1] and Chunjuan Liu[1]

[1]School of Electronic and Information Engineering, Lanzhou Jiaotong University, Lanzou 730070, China

[2]Corresponding author's e-mail: liyao@mail.lzjtu.cn

Abstract. In order to heighten their near infrared light-sensitive characteristics, lead phthalocyanine (PbPc) phototransistors were prepared by vacuum evaporation at various substrate temperatures (T_s) of 60°C, 100°C, 140°C and 180°C. The crystalline texture, absorption spectrum and surface morphology of PbPc films were measured. The light-sensitive characteristic measurement results indicated that the photoresponsivity (R) and specific detectivity (D^*) of the devices first rise and dwindle as T_s grows and a best performance is gained at T_s = 100°C. Research on the correlation between substrate-temperature-dependent film properties and device performance showed that the crystal texture, interconnection of grains, light absorption and carrier mobility of PbPc films could be varied by substrate heating, and thereby causing the variations in the device performance for R and D^*. Moreover, a model semi-quantitatively describing the variation of photocurrent with optical absorption coefficient and carrier mobility was set up, and some numerical calculations were made base on the model presented.

1. Introduction

Over the years, organic phototransistors (OPTs) have drawn more and more attention due to nice light absorption and mechanical flexibility of organic materials [1-3]. OPTs have potential applications in optical switch, photodetector and so on [3]. Behavior of OPTs relies greatly on quality of active layer, e.g., stacking mode [4], crystallinity [5] and grain interconnection [6,7]. And the quality of active layer is chiefly determined by preparation process (e.g., vacuum thermal evaporation and solution process). For a polycrystalline OPT prepared via vacuum thermal evaporation, the quality of active layer is inextricably associated with vacuum degree, evaporativity and substrate temperature (T_s). When the vacuum degree and evaporativity are certain, the quality of active layer is mostly influenced by T_s. It was proved that the preferential orientation of film growth could be altered via substrate heating [8,9].

Investigations on the correlation between lead phthalocyanine (PbPc) film properties and T_s have been made. For instance, Müller investigated the crystal texture changes of PbPc film with T_s [10]. Collins researched the dependence of the crystal texture and photoelectric properties of PbPc film on T_s [11,12]. But to date, the effects of T_s on the light-sensitive characteristics of PbPc OPTs have barely been described. Here, the PbPc OPTs were manufactured by vacuum thermal evaporation at various T_s, and the substrate-temperature-dependent performance of the devices were researched. And a semi-quantitative model that depicts the variation of photocurrent of OPTs with optical absorption coefficient and carrier mobility was built.

2. Experimental materials and methods

2.1. Active Layer Material

PbPc shows fine near infrared (NIR) absorption characteristics [13-16]. The crystal structures of PbPc species contain monoclinic and triclinic. It was demonstrated that the monoclinic PbPc film exhibits lower crystallinity than the triclinic [17]. The molecular stacking alone *c*-axis of monoclinic PbPc film restricts its carriers' transport in *ab* plane and brings low carrier mobility in *ab* plane [12]. Whereas, the molecular stacking alone *a*-axis of triclinic PbPc film enable it to exhibit superior transport property and carrier mobility in *ab* plane to the monoclinic [12]. Additionally, the NIR absorption of monoclinic PbPc film is inferior to that of the triclinic [17]. Consequently, generation of triclinic phase in PbPc film is profitable to NIR-sensitive PbPc OPTs.

2.2. Fabrication and Characterization

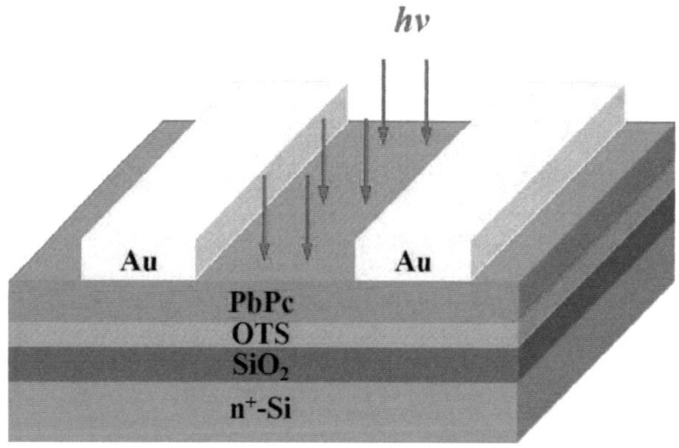

Figure 1. Structural graph of PbPc OPTs.

The structure of PbPc OPTs is shown in Figure 1. The devices were prepared according to the procedure reported in Ref [18]. The thickness of PbPc films was $t = 50$ nm, the vacuum pressure was maintained around 1.1×10^{-3} Pa and the deposition rate around 0.2 Å/s during evaporation, and the T_s were set to 60°C, 100°C, 140°C and 180°C, respectively. Au source and drain electrodes were evaporated onto PbPc films via a mask, and a channel length of 25 μm and width of 3 mm were achieved.

I-V characteristics were tested by semiconductor device test system. Absorption spectrum was tested by Lambda 950 spectrometer. Surface morphology was examined by atomic force microscopy (Agilent 7500), X-ray diffraction (XRD) patterns were measured by Rigaku D/Max-2400 X-ray diffractometer. Top irradiation was carried out by a laser diode with central wavelength at 880 nm.

3. Experimental results and discussion

3.1. Crystalline texture of PbPc film

The XRD patterns (Figure 2) display that the PbPc films consist of monoclinic and triclinic phases [19]. When $T_s \geqslant 100$ °C, as T_s grows, the intensity of the triclinic (100) line quickly rises and that of the triclinic (300) tardily rises whereas that of the monoclinic (200) fast dwindles. At the same time, the full width at half maximum of the triclinic (100) line is reduced from 0.32° to 0.16° as T_s elevates from 60°C to 180°C, indicating the preferential orientation of PbPc film growth is turned from monoclinic to triclinic and the crystallinity and content of the triclinic phase are both enhanced with T_s.

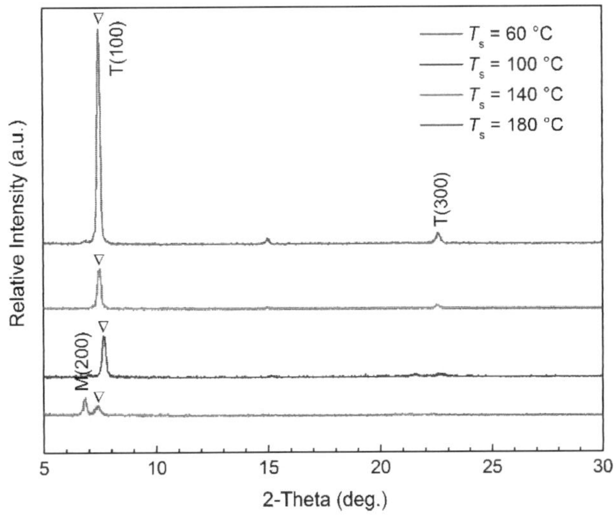

Figure 2. XRD patterns of PbPc films evaporated at various T_s.

3.2. Absorption spectrum of PbPc film

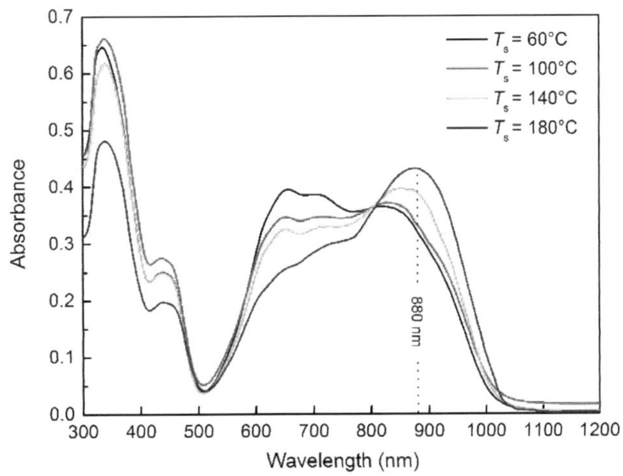

Figure 3. Absorption spectrum of PbPc films deposited on quartz glass at various T_s.

The absorption spectra (Figure 3) demonstrate that PbPc film has photo absorption in the ultraviolet, visible and NIR region. Particularly, its NIR absorption is attracting researchers' attention. We observed that the NIR absorption above 800 nm is augmented as T_s rises (see Figure 3). At 880 nm, the optical absorption coefficient (γ) of the films is increased with T_s. This may be attribute to the increment in the crystallinity and content of the triclinic phase of the films that is induced by substrate heating.

3.3. Surface morphology of PbPc film

Studies have found that transport property of organic film is related to its topography [6,20]. Carrier mobility of poly-crystalline organic film can be described as $1/\mu = 1/\mu_G + 1/\mu_B$ [21], here μ is carrier mobility, μ_B body mobility and μ_G grain boundary mobility. Grain boundary mobility can reflect block of grain boundaries to carrier transport [21] and is related to the grain interconnection that correlates with film surface rough degree. It has been suggested that, the bigger the surface rough degree, the worse the grain interconnection and the lower the grain boundary mobility [6].

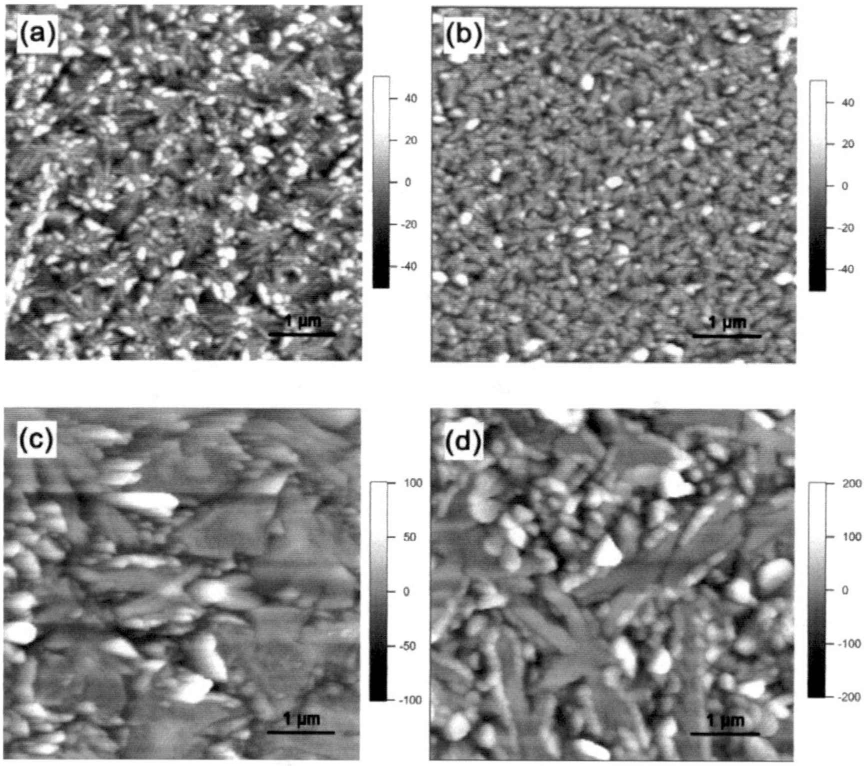

Figure 4. Atomic force microscopy images of PbPc films evaporated at (a) 60°C, (b) 100°C, (c) 140°C and (d) 180°C.

It can be seen from above, as T_s elevates from 60°C to 180°C, the increased crystallinity and content of the triclinic phase of PbPc films result in enhancement in μ_B of the films in *ab* plane. It is found from Figure 4, when T_s elevates from 60°C to 100°C, the root-mean-square roughness (δ) of the films is marginally decreased from 13.2 nm to 9.2 nm, and consequently the grain interconnection of them turns better and the μ_G of them gets greater. So, the enlarged μ_G and μ_B let the μ increase from 3.37×10^{-4} cm^2/V·s up to 6.86×10^{-4} cm^2/V·s (see Table 1). When T_s exceeds 140°C, the δ of the films increases quickly and reaches up to 45.1 nm, and the worse grain interconnection causes μ_G to decrease seriously. The increased μ_B and the sharply decreased μ_G finally yield a fast reduced μ from 5.68×10^{-4} cm^2/V·s down to 6.49×10^{-5} cm^2/V·s (see Table 1). It is concluded from above, when $T_s \geqslant$ 140 °C, the μ of the films may rely mainly on μ_G.

3.4. Performance of PbPc OPTs

Carrier mobility (μ) is OPTs' field effect parameter, and photoresponsivity (R) and specific detectivity (D^*) are OPTs' photosensitive parameters. Their definitions are seen Ref. [18,22]. As shown in Table 1, the PbPc OPTs manufactured at 60°C, 100°C, 140°C and 180°C were denoted as Dev. I-IV, respectively. The I-V characteristics (Figure 5) show that the PbPc OPTs demonstrate *p*-type field effect features. It can be seen from Table 1, as T_s grows, the I_{ph} (photocurrent), μ, D^* and R of the devices first rise and diminish and reach their maximum at T_s = 100 °C, at which I_{ph} = 392 nA, $\mu = 6.86 \times 10^{-4}$ cm^2/V·s, $D^* = 8.2 \times 10^8$ Jones and R = 12.16 mA/W. This can be attributed to substrate heating. The changes in the crystallinity and content of the triclinic phase and the grain interconnection of the films induced by substrate heating give rise to the variations in the photo absorption and carrier mobility, and thus bringing the changes in the device performance for I_{ph}, D^* and R. When T_s elevates from 60°C 100°C, the increment in the crystallinity and content of the triclinic phase and the optimization of grain interconnection of the films result in enhancement in the photo absorption and carrier mobility of the films, and hence leading to an increase in I_{ph}, D^* and R. When T_s = 100 °C, a best device performance is realized. When T_s exceeds 100°C, the quick degradation of grain interconnection of the films causes carrier mobility to reduce greatly, and the fast decreasing carrier mobility and the slowly increasing photo absorption finally produces a lower I_{ph}, D^* and R.

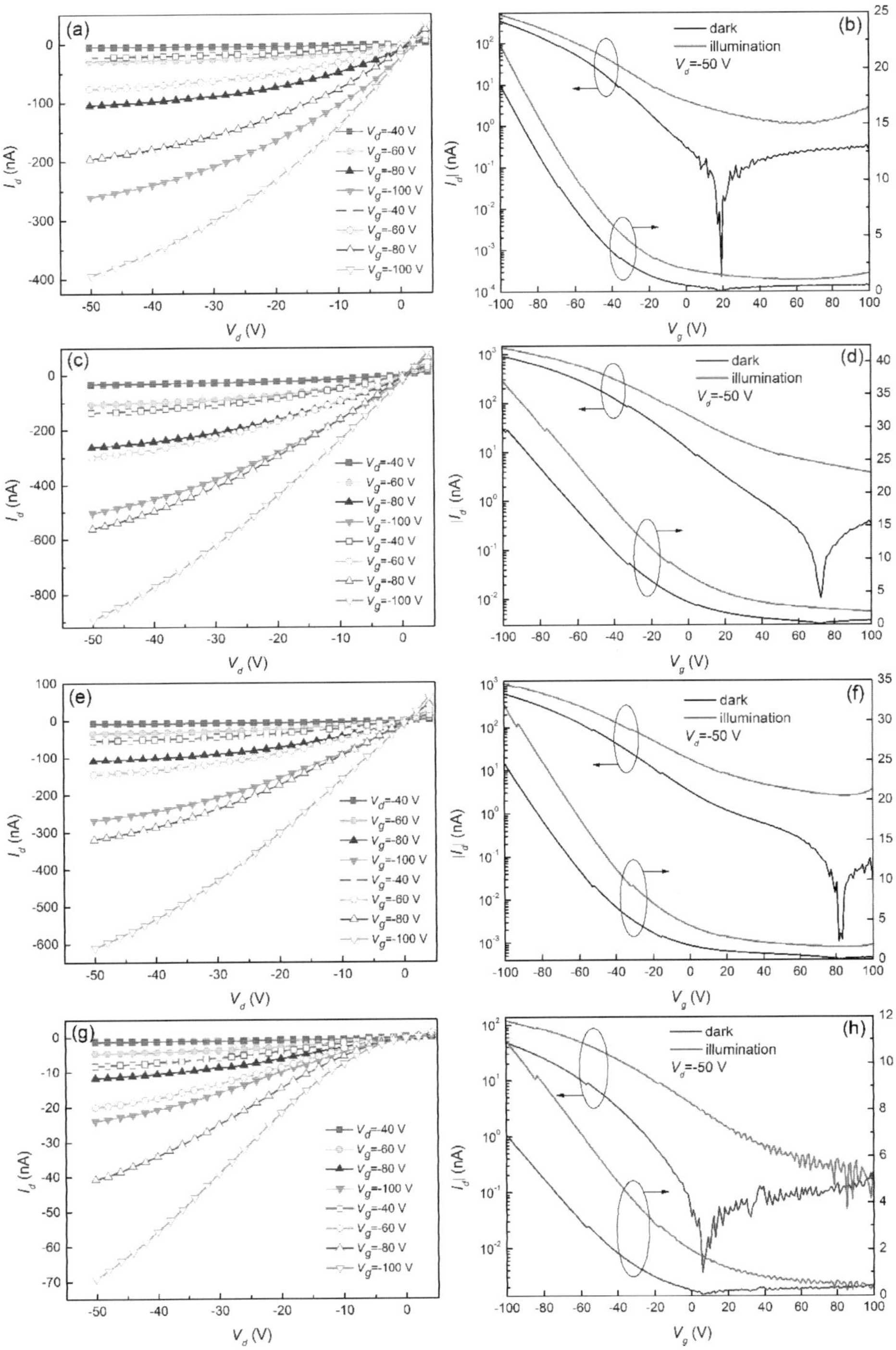

Figure 5. I-V characteristics of Dev. I (a and b), Dev. II (c and d), Dev. III (e and f) and Dev. IV (g and h).

Table 1. Performance parameters of PbPc OPTs manufactured at various T_s.

Devices No.	T_s (°C)	I_{ph} (nA)	μ (cm²/V·s)	R (mA/W)	D^* (Jones)	δ (nm)
I	60	130	3.37×10^{-4}	3.58	3.5×10^8	13.2
II	100	392	6.86×10^{-4}	12.16	8.2×10^8	9.2
III	140	345	5.68×10^{-4}	8.77	8.0×10^8	18.6
IV	180	42	6.49×10^{-5}	1.02	3.7×10^8	45.1

3.5. Model for Drift Photocurrent

The photoelectric conversion process occurring in *p*-type OPTs contains (i) exciton generation; (ii) exciton diffusion; (iii) exciton separation; (iv) collection of photo-generated holes (PGHs) by drain. Through this process, incident light signal is transformed into electric signal. In this place, presuming that (i) the conversion efficiency of photon to photo-produced carrier is invariant; and (ii) the photocurrent of OPTs is drift current, which could be simply denoted as [20]

$$I_{ph} = A \cdot qp\mu E . \tag{1}$$

where A is cross sectional area of OPTs, q elementary charge, p density of PGHs, and $E \approx V_{ds}/L$ is channel electric field (V_{ds}, drain-source voltage). The relation between the densities of the PGHs and the photons absorbed by active layer can be expressed as [23]

$$p \propto N = \frac{P_{inc}\left(1-e^{-\gamma t}\right)}{h\nu} . \tag{2}$$

where P_{inc} is incident optical power, N density of photons, $h = 6.62 \times 10^{-34}$ J·s, $t = 50$ nm (ditto), and ν photon frequency. Combining equation (1) and (2) gives

$$I_{ph} \propto N = \frac{P_{inc}\left(1-e^{-\gamma t}\right)}{h\nu} . \tag{3}$$

Equation (3) means that photocurrent shows a positive correlation with optical absorption coefficient.

Here, some numerical computations were made according to the model and test data. We let N_j and γ_j (j = I, II, III and IV) represent the density of photons and the optical absorption coefficient of active layer, separately. Thus, $\gamma_I = 6.38 \times 10^4$ cm^{-1}, $\gamma_{II} = 6.57 \times 10^4$ cm^{-1}, $\gamma_{III} = 7.79 \times 10^4$ cm^{-1}, and $\gamma_{IV} = 8.58 \times 10^4$ cm^{-1}. And we let $I_{ph\_j\_theor}$, p_j and μ_j (j = I, II, III and IV) denote the theoretical photocurrent, the density of PGHs and the field effect mobility, separately. Then $\mu_I = 3.37 \times 10^{-4}$ cm²/V·s, $\mu_{II} = 6.86 \times 10^{-4}$ cm²/V·s, $\mu_{III} = 5.68 \times 10^{-4}$ cm²/V·s, and $\mu_{IV} = 6.49 \times 10^{-5}$ cm²/V·s. Combining equation (1), (2) and (3) gives

$$\begin{aligned} \frac{I_{ph\_II\_thero}}{I_{ph\_I\_thero}} &= \frac{A \cdot qp_{II}\mu_{II}E}{A \cdot qp_I\mu_I E} = \frac{p_{II}\mu_{II}}{p_I\mu_I} = \frac{N_{II}\mu_{II}}{N_I\mu_I} \\ &= \frac{\frac{P_{inc}\left(1-e^{-\gamma_{II}t}\right)}{h\nu}\cdot\mu_{II}}{\frac{P_{inc}\left(1-e^{-\gamma_I t}\right)}{h\nu}\cdot\mu_I} = \frac{\left(1-e^{-\gamma_{II}t}\right)\cdot\mu_{II}}{\left(1-e^{-\gamma_I t}\right)\cdot\mu_I} . \\ &\approx 2.08 \end{aligned} \tag{4}$$

Similarly, we get $\frac{I_{ph\_III\_thero}}{I_{ph\_I\_thero}} = \frac{\left(1-e^{-\gamma_{III}t}\right)\cdot\mu_{III}}{\left(1-e^{-\gamma_I t}\right)\cdot\mu_I} \approx 1.99$ and $\frac{I_{ph\_IV\_thero}}{I_{ph\_I\_thero}} = \frac{\left(1-e^{-\gamma_{IV}t}\right)\cdot\mu_{IV}}{\left(1-e^{-\gamma_I t}\right)\cdot\mu_I} \approx 0.25$. And on the basis of the test data, the photocurrent of Dev. I is 130 nA, denoted as $I_{ph\_I\_test}$ = 130 nA. Then, we have $I_{ph\_II\_test}$ = 392 nA, $I_{ph\_III\_test}$ = 345 nA and $I_{ph\_IV\_test}$ = 42 nA, and $\frac{I_{ph\_II\_test}}{I_{ph\_I\_test}} \approx 3.01$, $\frac{I_{ph\_III\_test}}{I_{ph\_I\_test}} \approx 2.65$ and $\frac{I_{ph\_IV\_test}}{I_{ph\_I\_test}} \approx 0.32$.

Table 2. The computed results based on the model presented and test data.

Model computation values		Experiment computation values	
$\dfrac{I_{ph\_II\_theor}}{I_{ph\_I\_theor}}$	2.08	$\dfrac{I_{ph\_II\_test}}{I_{ph\_I\_test}}$	3.01
$\dfrac{I_{ph\_III\_theor}}{I_{ph\_I\_theor}}$	1.99	$\dfrac{I_{ph\_III\_test}}{I_{ph\_I\_test}}$	2.65
$\dfrac{I_{ph\_IV\_theor}}{I_{ph\_I\_theor}}$	0.25	$\dfrac{I_{ph\_IV\_test}}{I_{ph\_I\_test}}$	0.32

The values of $\frac{I_{ph\_j\_thero}}{I_{ph\_I\_thero}}$ and $\frac{I_{ph\_j\_test}}{I_{ph\_I\_test}}$ were calculated and compared (Table 2), in which j = II, III and IV denote Dev. II, III and IV, separately. The computed results indicate that the change trend of the two sets of the calculated values is the same, however, there are still errors between them. The same change trend of them demonstrates the validity of the model presented, and the existing errors indicate that the precision of the mode is insufficient and needs to be improved.

4. Conclusions

The PbPc OPTs were manufactured at various T_s. The photosensitive characteristic measurement results showed, as T_s elevates, the R and D^* of the devices first rise and diminish and a best performance for R and D^* is gained at T_s = 100°C. Surveys on the correlation between substrate-temperature-dependent film properties and device performance suggested that, the changes in the crystallinity and content of triclinic phase and the grain interconnection of the films produced by substrate heating bring the variations in the optical absorption and carrier mobility of the films, and thus leading to the changes in the device performance for I_{ph}, R and D^*. This study can provide an effective route to improve the performance of PbPc OPTs. Besides, a semi-quantitative model was set up and can be used to depict variations of photocurrent with optical absorption coefficient and carrier mobility.

Acknowledgments

This work was supported by Tianyou Youth Talent Lift Program of Lanzhou Jiaotong University.

References

[1] P. C. Y. Chow, N. Matsuhisa, P. Zalar, M. Koizumi, T. Yokota and T. Someya, 2018, Dual-gate organic phototransistor with high-gain and linear photoresponse, *Nat. Commun.*, **9**, 4564.

[2] F.-J. Kahle, T. Hahn, T. Mukhopadhyay, B. Puttaraju, S. Patil, A. Köhler, 2019, Organic Bidirectional Phototransistors Based on Diketopyrrolopyrrole and Fullerene, *Adv. Funct. Mater.*, **29**, 21, 1805684.

[3] W. T. Huang, Y.-H. Lin and T. D. Anthopoulos, 2018, High Speed Ultraviolet Phototransistors Based on an Ambipolar Fullerene Derivative, *ACS Appl. Mater. Interfaces*, **10**, 12, 10202-10210.

[4] X. Kong, X. Zhang, D. M. Gao, D. D. Qi, Y. L. Chen and J. Z. Jiang, 2015, Air-stable ambipolar field-effect transistor based on a solution-processed octanaphthoxy-substituted tris (phthalocyaninato) europium semiconductor with high and balanced carrier mobilities, *Chem. Sci.*, **6**, 3, 1967-1972.

[5] Gianluca Generali, F. Dinelli, R. Capelli, S. Toffanin, F. di Maria, M. Gazzano, G. Barbarella, M. Muccini, 2011, Correlation among morphology, crystallinity, and charge mobility in OFETs made of quaterthiophene alkyl derivatives on a transparent substrate platform, *J. Phys. Chem. C*, **115**, 46, 23164-23169.

[6] L. Jason and Z. N. Bao. 2006, Effect of morphology on organic thin film transistor sensors, *Anal. Bioanal. Chem.*, **384**, 2, 336-342.
[7] H. M. Lee, H. Moon, H.-S. Kim, Y. N. Kim, S.-M. Choi, S. Yoo, S. O. Cho, 2011, Abrupt heating-induced high-quality crystalline rubrene thin films for organic thin-film transistors, *Org. Electron.*, **12**, 8, 1446-1453.
[8] F. Medjani, R. Sanjinés, G.Allidi, A.Karimi, 2006, Effect of substrate temperature and bias voltage on the crystallite orientation in RF magnetron sputtered AlN thin films, *Thin Solid Films* **515**, 1, 260-265.
[9] C. Schünemann, C. Elschner, A. A. Levin, M. Levichkova, K. Leo, M. Riede, 2011, Zinc phthalocyanine-Influence of substrate temperature, film thickness, and kind of substrate on the morphology, *Thin Solid Films*, **519**, 11, 3939-3945.
[10] C. Hamann, A. Mrwa, M. Müller, 1991, Lead phthalocyanine thin films for NO_2 sensors, *Sensors and Actuators B*, **4**, 1-2, 73-78.
[11] R. A. Collins, A. Krier and A. K. Abass, 1993, Optical properties of lead phthalocyanine (PbPc) thin films, *Thin Solid Films*, **229**, 1, 113-118.
[12] R. A. Collins, A. Belghachi, 1989, Structural properties of lead phthalocyanine thin films, *Mater. Lett.*, **8**, 9, 349-352.
[13] X. Wang, H. F. Li, Z. S. Su, F. Fang, G. Zhang, J. B. Wang, B. Chu, X. Fang, Z. P. Wei, B. Li, W. L. Li, 2014, Efficient organic near-infrared photodetectors based on lead phthalocyanine/C_{60} heterojunction, *Org. Electron.*, **15**, 10, 2367-2371.
[14] M.S. Choi, S. Lee, H.J. Kim, J.J. Kim, 2018, Inverted near-infrared organic photodetector with oriented lead (II) phthalocyanine molecules via substrate heating, *Org. Electron.*, **61**, 164-169.
[15] J. P. Mudrick, W. R. Cao, J. Li, J. G. Xue, 2014, Tandem small molecule organic photovoltaic cells with broad spectral response up to 1 μm and a high open-circuit voltage, *Org. Electron.*, **15**, 11, 3024-3030.
[16] K. Vasseur, K. Broch, A. L. Ayzner, B. P. Rand, D. Cheyns, C. Frank, F. Schreiber, M. F. Toney, L. Froyen, P. Heremans, 2013, Controlling the texture and crystallinity of evaporated lead phthalocyanine thin films for near-infrared sensitive solar cells, *ACS Appl. Mater. Interfaces*, **5**, 17, 8505-8515.
[17] H.-S. Shim, H. J. Kim, J. W. Kim, S.-Y. Kim, W.-I. Jeong, T.-M. Kim and J.-J. Kim, 2012, Enhancement of near-infrared absorption with high fill factor in lead phthalocyanine-based organic solar cells, *J. Mater. Chem.* **22**, 18, 9077-9081.
[18] L. Du, X. Luo, F. Zhao, W. Lv, J. Zhang, Y. Peng, Y. Tang, Y. Wang, 2016, Toward facile broadband high photoresponse of fullerene based phototransistor from the ultraviolet to the near -infrared region, *Carbon*, **96**, 685-694.
[19] A. Miyamoto, K. Nichogi, A. Taomoto, T. Nambu, M. Murakami, 1995, Structural control of evaporated lead-phthalocyanine films, *Thin Solid Films*, **256**, 1-2, 64-67.
[20] D. Tu, L. Herlogsson, L. Kergoat, X. Crispin, M. Berggren, R. Forchheimer, 2011, A Static Model for Electrolyte-Gated Organic Field-Effect Transistors, *IEEE Trans. Electron Devices*, **58**, 10, 3574-3582.
[21] J. H. Schön, C. Kloc, B. Batlogg, 2000, On the intrinsic limits of pentacene field-effect transistors, *Org. Electron.*, **1**, 1, 57-64.
[22] X. Luo, Y. Li, W. Lv, F. Zhao, L. Sun, Y. Peng, Z. Wen, J. Zhong and J. Zhang, 2015, Position-dependent performance of copper phthalocyanine based field-effect transistors by gold nanoparticles modification, *Nanotechnology*, **26**, 035201.
[23] T. M. Clarke and J. R. Durrant, 2010, Charge Photogeneration in Organic Solar Cells, *Chem. Rev.*, **110**, 6736-6767.

Effects of Ag/Ta co-implanted TiN coating on medical titanium alloy on biological and antibacterial properties

Xiduo Song[1], Mengli Zhao[1,2] and Dejun Li[1,3]

[1]College of Physics and Materials Science, Tianjin Normal University, Tianjin 300387, China

[2]Corresponding Author's E-mail: zhaomengli_56@163.com
[3]Corresponding Author's E-mail: dli1961@126.com

Abstract. In this work, co-implantation of silver and tantalum on TiN coated Ti-6Al-4V surface was carried out by an omnidirectional ion implantation system. The effects of Ag/Ta co-implantation on its biological properties and antibacterial activity were studied. The elemental content was determined by XPS, the biocompatibility was determined by Fibroblast-like cells (L929) and the antimicrobial activity was determined by *Escherichia coli* (*E. coli*). The results indicated that Ag/Ta-TiN coating exhibited the good biocompatibility and excellent antibacterial properties, which means that Ag/Ta co-implantation can improve the biocompatibility and antibacterial properties of Ti-6Al-4V.

1. Introduction

Osteonecrosis is a common disease in orthopedic department. Artificial joint replacement is an ideal treatment method in clinic. Titanium alloy (Ti-6Al-4V), as a commonly used artificial joint, has been widely used in orthopedic implants due to its good biocompatibility and processability. However, medical titanium alloy has the disadvantages of poor antibacterial activity and biological inertia [1-4], which makes it prone to material degradation and bacterial infection after surgical implantation, and finally leads to surgical failure, which seriously hinders its development prospects in hard tissue repair. Therefore, it is imperative to modify titanium alloy to enhance its antibacterial.

On the other hand, ideal orthopedic biomaterials need good biocompatibility. In recent years, tantalum (Ta) has attracted more attention in the medical field because of its good biocompatibility [5-8], non-toxic and non-irritating to human body, excellent corrosion resistance and other advantages. Therefore, Ta can be combined with Ti-6Al-4V to improve its biocompatibility. Good bone replacement materials should also have excellent antimicrobial properties. However, the antibacterial mechanism of Ta is still not clear, so it is crucial to find another material that can combine with Ta to improve the antibacterial properties of Ti-6Al-4V. silver (Ag) has good long-term antibacterial properties and can kill bacteria at very low concentrations [9-12]. It is often used in the field of biomedical science, so Ag combine with Ta co-implantation maybe improve the antibacterial properties of Ti-6Al-4V.

In this paper, a layer of TiN was deposited on the titanium alloy, which can not only effectively prevent the release of toxic ions (V^{4+}, Al^{3+}) into the human body, but also improve the hardness of the material. And Ag and Ta were co-implanted into TiN to explore its biological and antimicrobial activities through cell proliferation and antibacterial experiments.

2. Material and Methods

2.1. Preparation of Ag/Ta-TiN
Ti-6Al-4V wafers (t=2 mm, d=10 mm) and silicon (100) wafers (Ti-6Al-4V alloys are used for *in vitro* biological experiments, and silicon wafers are used to measure the number of elements) were selected as substrates. Beforeusing, wash them in distilled water and anhydrous ethanol for 20 min, 3 times each, and then dry them for use. First, the JGP-450 magnetron co-sputtering system was used to deposit TiN film. The substrate was first put into the chamber, and the chamber was vacuumed to 5×10^{-4} Pa, then Ar (purity 99.999 %) was passed into the chamber. When the air pressure changed to 5 Pa, the substrate was pre-sputtering cleaned by - 600V pulse bias for 10 min. After cleaning, N_2 was added (The ratio of N_2 to Ar is 1:12), the air pressure was adjusted to 0.5 Pa, the power of RF power was 100 W, and the sputtering time was 60 min.

Secondly, plasma immersion ion implantation system (PIII & D-700, China.) was used to co-implantation Ag and Ta into TiN. Ag and Ta targets with purity of 99.999 % and 99.999 % were used as sources. Experimental parameters and detailed sample preparation process can be found in our previous work [13].

2.2. Structure and composition analysis
The chemical composition of the samples was analyzed by X-ray photoelectron spectroscopy (XPS, PHI5000, VersaProbe.).

2.3. In vitro cell experiment
Fibroblast-like cells (L929) were used as the target cells in this experiment. Cells were cultured in complete α-minimum essential medium (α-MEM, Gibco.) containing 10 % fetal bovine serum (FBS, BI) and 1% anti-microbial penicillin and streptomycin (Hyclone) in an incubator at 37 ℃ with 5 % CO_2 and saturated humidity. The medium was changed daily and only the 3-5 generation cells were used in the experiment. Cells were inoculated into the samples at a density of 2×10^4 cells/mL to evaluate the cellular compatibility of the samples.

After 1, 2, 3, and 4 days of culture, cell proliferation was assessed by cell counting kit-8 (CCK-8, Genview.) analysis. 30 μL CCK-8 and 270 μL medium were added to each sample and cultured at 37 ℃ and 5 % carbon dioxide for 2 h. Finally, transfer the liquid from each well to a 96-well plate and divide into 3 Wells, averaging 100 μL per well. The optical density (OD) value of each sample was recorded at 450 nm using an enzyme linked immunoassay (EK800, BioTek.) with three measurements.

After 4 days of culture, they were rinsed with phosphate buffer saline (PBS, pH 7.4, HyClone.) 3 times, 5 min each, and then fixed with 2.5 % glutaraldehyde for 4-8 h. The samples were dehydrated successively in gradient ethanol solutions (30, 50, 70, 90 and 100 V/V %). Then, ethanol and tert-butanol were used to dehydrate the cells again at ratios of 1:3, 2:2, 3:1 and 1:0 for 15 min each time. The dehydrated samples were frozen in 1 mL tert-butyl alcohol. After vacuum drying, the cell morphology of the samples was observed by scanning electron microscopy (SEM, SU-8010, Hitachi.) under 5 kV voltage.

2.4. Antibacterial assessment
The experimental strains were *Escherichia coli (E. coli*), and the experimental procedures were as follows: (1) Put *E. coli* into LB liquid medium and oscillate in a thermostatic oscillator for 2-3 h (temperature controlled at 37 ℃). The absorbance photometric value (OD) of the shaken suspension was determined to be 0.5, and the bacterial suspension was obtained by diluting 10^3 times with PBS buffer solution for later use. (2) Add 1 mL of the prepared bacterial suspension to the sterilized and dried samples, and culture it in a mold incubator for 24 h (37 ℃). (3) After 24 h, the bacterial liquid after interaction with the sample was mixed with 10 mL PBS and ultrasonic vibration was carried out for 5 min. 100 μL bacterial liquid was absorbed for coating plate, and cultured for 24 h, then photographed and counted, and observed results. (4) Take 3 samples of each material, and repeat the

experiment for 3 times for each sample, and take the average value. Then the bacteriostatic rate can be calculated. After 24 h culture with *E. coli*, the surface morphology of bacteria was observed by SEM.

3. Results and Discussion

3.1. Composition analysis
The chemical states of TiN and Ag/Ta-TiN samples were analyzed by XPS, and the results are shown in Figure 1(a), and the atomic percentage is shown in Table 1. It can be seen from the full spectrum that Ag and Ta peaks appear in the Ag/Ta-TiN sample after ion implantation.

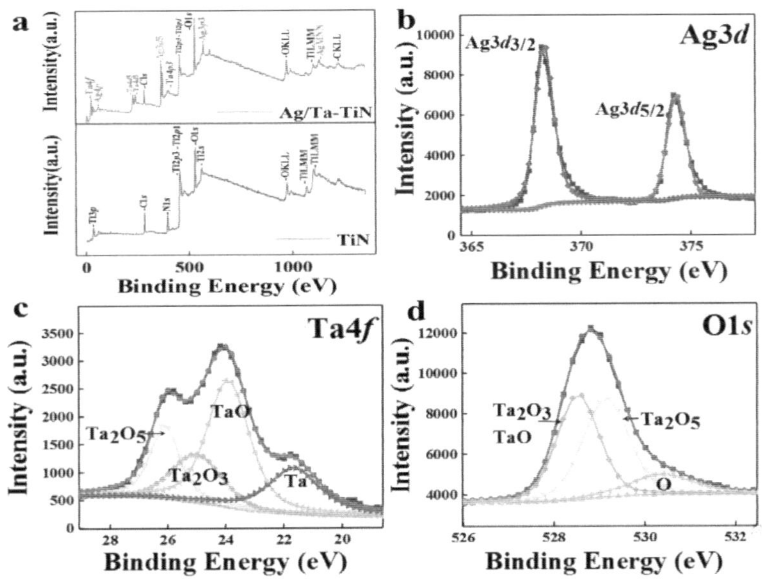

Figure 1. (a) XPS investigation spectra of the sample; (b) high resolution Ag3*d* XPS spectra; (c)high resolution Ta4*f* XPS spectra, and (d) high resolution O1*s* XPS spectra.

As shown in Figure 1(b-d), in order to further study the surface composition of the samples implanted with Ag and Ta, the high-resolution XPS spectra of Ag3*d*, Ta4*f* and O1*s* are analyzed.

In the Ag3*d* spectrum, the two main peaks of 374.05 eV ($3d_{3/2}$) and 368.04 eV ($3d_{5/2}$) indicate that the implanted Ag ions exist in a metallic state [14]. In the Ta4*f* spectrum, the four main peaks of 25.9 eV (Ta_2O_5), 25.0 eV (Ta_2O_3), 23.8 eV (TaO) and 21.9 eV (Ta) indicate that most of the implanted Ta ions combine with oxygen to form oxides [15]. Meanwhile, Ta_2O_5 peaks can be observed in the O1*s* spectrum of 530.2 eV, and Ta_2O_3 and TaO peaks can be observed in the O1*s* spectrum of 528.0 eV [15], which also confirm the view that Ta mainly exists in oxides.

Table 1. Atomic percent of Ag/Ta-TiN.

Experiment groups	Relative percent of Ag3*d* (%)	Relative percent of Ta4*f* (%)
Ag/Ta-TiN	5.14	2.96

3.2. Cytocompatibility
The proliferation of L929 cells on the samples was measured by CCK-8 method and tested after 1, 2, 3, and 4 days of culture, respectively. The results are shown in Figure 2(a). It can be seen that after 2 days of culture, there is basically no difference in the number of cells on the surface of the three

IOP Conf. Series: Materials Science and Engineering 1167 (2021) 012011 doi:10.1088/1757-899X/1167/1/012011

groups of samples, mainly because the lower planting density and the proliferation of cells needed a period of time. After 4 days of culture, the number of Ag/Ta-TiN cells on the surface of the samples is significantly higher than that of Ti-6Al-4V, indicating that the implantation of Ta improved the proliferation ability of cells.

In order to further explore the growth of L929 cells on the sample surface, SEM was used to observe the sample surface. The results are shown in Figure 2(b and c). The number of cells on Ag/Ta-TiN samples co-implanted is significantly more than that on Ti-6Al-4V, which is also the same as the results of CCK-8 experiment. Most of the cells on Ti-6Al-4V are in a single state, and some cells are spherical, but not fully expanded. The cells on the surface of the Ag/Ta-TiN samples are seen to have partially broken out of the sphere and began to spread out, and some cells are clearly in the mitotic stage. However, the cells on the Ti-6Al-4V samples show no signs of division at all, which is sufficient to indicate that the samples after co-implantation are more suitable for cell initial adhesion and diffusion and more conducive to cell growth.

In conclusion, the implantation of Ag/Ta can promote cell adhesion and diffusion, and promote cell proliferation.

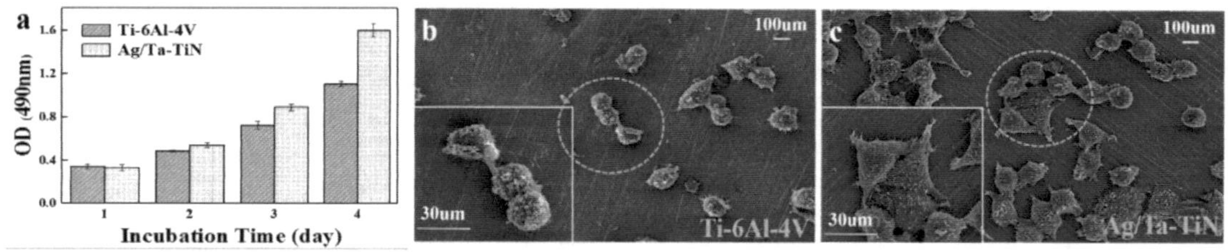

Figure 2. (a)The histogram of L929 cell proliferation was observed after 1, 2, 3 and 4 days of culture; SEM images of L929 cells on the surfaces of (b) Ti-6Al-4V; (c) Ag/Ta-TiN.

3.3. *In vitro antibacterial assay*

In order to further evaluate the antibacterial properties of the samples, the experiment was carried out with *E. coli*, and the adherent bacteria were separated from the samples and re-cultured on AGAR plate. As shown in Figure 3(a), the bacterial survival rate of the control group (Ti-6Al-4V) was 100% as the standard, and the bacteriostatic rate of the Ag/Ta-TiN film surface is 97.3%. The growth of the colony on the AGAR plate is shown in Figure 3(b and c). It can be seen that compared with the isolate and reculture colonies on Ti-6Al-4V, the colonies on Ag/Ta-TiN are significantly reduced, which indicate that the implantation of Ag/Ta significantly improve the antibacterial performance of the sample.

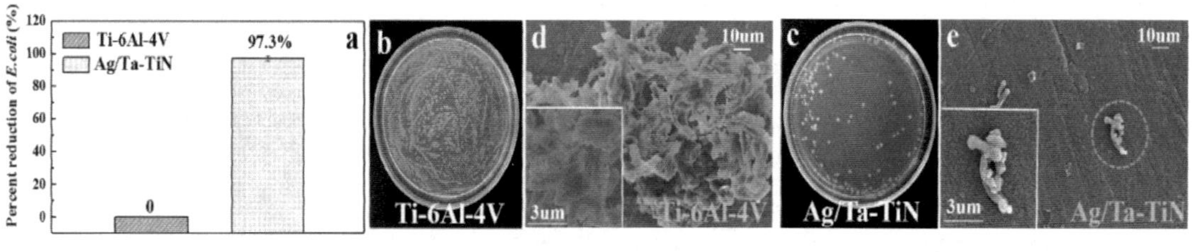

Figure 3. (a) Ti-6Al-4V and Ag/Ta-TiN antibacterial rate of samples; Recultured *E. coli* colonies on agar plates of (b) Ti-6Al-4V; (c) Ag/Ta-TiN; SEM images of *E. coli.* on (d) Ti-6Al-4V; (e) Ag/Ta-TiN.

SEM was used to further study the morphology of bacteria on the sample surface, as shown in Figure 3(d and e). It can be find that the number of bacteria on theTi-6Al-4V sample is the largest, and the morphology is intact and rod-like, and most of them gather together in the form of clusters, which indicates that the Ti-6Al-4V sample basically does not have the ability to Fight bacteria. However, the number of bacteria on the samples after Ta/Ag co-implantationis significantly reduced, and the morphology of bacteria is obviously damaged. Most of the bacteria are distorted and corrugated, and even some are broken, indicating that bacteria could not survive on the surface basically. This is consistent with the results in Figure 3(a).

This good antibacterial activity may be because Ag has good antibacterial properties, which causes more damage to the structure of the *E.coli* membrane and reduces the activity of membrane enzymes [16], thus killing the bacteria.

4. Conclusion

In this paper, TiN coating was prepared on the surface of Ti-6Al-4V, and then Ag and Ta were co-implanted into the TiN coating. The effects of Ag and Ta on the biological properties and antibacterial activity of TiN coating were studied. The results showed that the co-implanted Ag/Ta-TiN was superior to Ti-6Al-4V in cell proliferation and initial antimicrobial activity. Ta ion implantation can promote cell proliferation and differentiation, and Ag ion can improve the antibacterial properties of the sample.

In conclusion, co-implantation Ag/Ta can effectively improve the cytocompatibility and antibacterial activity of Ti-6Al-4V coated by TiN. In general, This work sheds light on the application of titanium alloy in orthopedics.

Acknowledgments

This work was supported by National Natural Science Foundation of China (51772209), Natural Science Foundation of Tianjin (No.18JCQNJC72000), and Program for Innovative Research in University of Tianjin (Grant No. TD13-5077) and The Post-graduate and doctoral Innovation Project of Tianjin (2019YJSS137).

References

[1] Zhao Y, Cao H, Cheng T, Qian S and Chu P K 2015 *ACS Appl. Mater. Interfaces* **7** 17826–17836
[2] Behera R R, Das A, Hasan A, PamuD, Pandey L M and Sankar M R 2020 *J. Alloys Compd.* **842** 155683
[3] Silva D, *et al* 2020 *Bioelectrochem. Bioenerg.* **136** 107622
[4] Kazemi M, Ahangarani S, Esmailian M and Shanaghi A 2020 *Surf. Coat. Technol.* **397** 126044
[5] Hu X, Mei S,Wang F, Qian J, Xie D, Zhao J, Yang L L, Wu Z Y and Wei J 2021 *Bioact. Mater* **6** 928–940
[6] Bermúdez M-D, Carrión F J, Martínez-Nicolás G and López R 2005 *J. Frict. Wear.* **258** 693–700
[7] Dou X, Wei X, Liu G, Wang S, Lv Y, Li J and ZhaoD 2019 J. *Orthop. Translat.* **19** 81-93
[8] Wu S, *et al* 2021 *Mater. Sci. Eng. C* **120** 111777
[9] Necula B S, Fratila-Apachitei L E, Apachitei I and Duszczyk J 2009 *Acta Biomater***5** 3573–3580
[10] Cao H, Liu X, Meng F and Chu P K 2011 *Biomaterials* **32** 693–705
[11] Huang D, Lin Q X, Zhou Y J, Li J, Wei Y, Hu Y C, Lian X J, Chen S and Chen W Y 2020 *Surf. Coat. Technol.* **399** 126169
[12] Devi A P, Padhi D K, Mishra P M and Behera A K 2021 *J. Environ. Chem. Eng.* **9** 104778
[13] Han X, Ji X, Zhao M and Li D 2020 *Surf. Coat. Technol* **397** 126020
[14] Li J, Qiao Y, Ding Z and Liu X 2011 *Surf. Coat. Technol* **205** 5430-5436
[15] McNamara K, Belochapkine S, Hossain K and Tofail S A 2021 *Appl. Surf. Sci* **535** 147621
[16] Calabrese G, Petralia S, Franco D, Nocito G, Fabbi C, Forte L, Guglielmino S, Squarzoni S, Traina Fand Conoci S 2021 *Mater. Sci. Eng, C* **118** 111394

Effect of chemical copolymerization and mixed chain extenders on mechanical properties of HTPB polyurethane

Yu Zhang[1], Jian Zheng[1,5], Xiao Zhang[2,6], Yongqiang Du[3], Ke Li[4], Guibo Yu[1], Yunfei Jia[1] and Yahao Liu[1]

[1]Shijiazhuang Campus, Army Engineering University, Shijiazhuang 050003, China
[2]Engineering University of PAP, Xi'an 710086, China
[3]Dalian Naval Academy, Dalian 116013, China
[4]College of Naval Architecture and Ocean Engineering, Naval University of Engineering, Wuhan 430033, China

[5]Corresponding Author's E-mail: zhengj2020@163.com
[6]Corresponding Author's E-mail: zxleo@foxmail.com

Abstract. In this study, three kinds of polyurethane materials based on HTPB were prepared by copolymerization of hydroxyl-terminated polybutadiene (HTPB) and hydroxyl-terminated polyethylene glycol-tetrahydrofuran copolyether (HTPE) and adding mixed chain extender, polyether polyols HEP-330N and 2-hydroxyethyl disulfide (HEDS). The effects of copolymerization and mixed chain extender on the mechanical properties of polyurethane were studied by static tensile test, dynamic mechanical property analysis (DMA) and differential scanning calorimetry (DSC). Among them, compared with pure HTPB, the tensile strength and elongation at break of polyurethanes obtained by copolymerization, the polyurethanes obtained by adding mixed chain extender, and the polyurethanes obtained by both of these two methods were significantly improved, and the storage modulus and loss modulus were also improved. In addition, the glass transition temperature of polyurethane remained basically unchanged, which was in the range of -79℃ to -75℃. It was much lower than the room temperature, which would make it more widely used. This effective method will provide an opportunity for the synthesis of various high-performance composites.

1. Introduction

Hydroxyl-terminated polybutadiene (HTPB) is a kind of low molecular weight liquid rubber [1,2], which has the characteristics of good transparency, low viscosity, low glass transition temperature and easy processing [3-8]. Therefore, the materials prepared from HTPB have excellent mechanical properties. In addition, it is resistant to water, oil-resistant, acid-resistant corrosion, and electrical insulation performance [9], which is generally applied to elastomers, fibers, foams, binders, and the like [8]. However, with the more and more extensive application of HTPB, the requirements for its comprehensive properties become higher and higher, so there is an urgent need to further improve the mechanical properties of HTPB polyurethane [8].

Chemical copolymerization is a simple, common and effective method. The crystallization of the polymer during curing can be reduced by copolymerization, thus reducing the hindrance to the molecular chain. Moreover, because of the designability of the composites, the copolymerization can also introduce the properties that they did not have before [10,11]. Hydroxy-terminated polyether

(HTPE) is a block copolymer of polyethylene glycol and polyteenhydrofuran [12,13]. Its molecular main chain contains a large number of ether bonds and no side chain groups, so the glass transition temperature is very low, the molecular chain flexibility is good, and its viscosity is low and the rheological property is good [13,14]. Therefore, HTPE and HTPB were used for copolymerization in this study.

Polyurethane is generally formed by the polymerization of long-chain polyols and diisocyanates, long-chain polyols are called soft segments, and diisocyanates are called hard segments. Among them, the soft segment can provide elasticity and toughness for polyurethane, while the hard segment can provide strength and modulus for polyurethane [15-17]. Further, the hard segment also includes chain extender and cross-linking agent, which play the role of increasing the length of the polymer chain and increasing the crosslink density, respectively. Polyether polyols HEP-330N is not only a kind of macromolecular polyol, but also a macromolecular chain extension cross-linking agent. The difference between polyether polyols HEP-330N and general macromolecular polyol is that 330N molecular chain contains a large number of active primary hydroxyl groups, which can increase the reaction activity of the system and accelerate the chain extension reaction of the system, and the three hydroxyl groups can form cross-linking points between the molecular chains of elastomers. Therefore, it can not only act as a soft segment structure but also play a role of chain extension and cross-linking in polyurethane elastomers [18,19]. Therefore, this study used the combination of macromolecular chain extender 330N and small molecular chain extender hydroxyethyl disulfide (HEDS) to form a mixed chain extender to improve the flexibility, strength and modulus of polyurethane at the same time.

In a word, the method of improving the mechanical properties of HTPB polyurethane by copolymerization and adding mixed chain extender was reported in this paper. The modified polyurethane was prepared and characterized, and the effect of copolymerization and mixed chain extender was studied.

2. Materials and experiments

2.1. Materials

Hydroxyl-terminated polybutadiene (HTPB, 99%, molecular weight 2880, hydroxyl value 0.75mmol/g) and hydroxyl-terminated polyethylene glycol tetrahydrofuran copolyether (HTPE, molecular weight 3000, hydroxyl value 0.68mmol/g) were provided by Liming Research&Design Institute of Chemical Industry Co., Ltd., Luoyang, China. Polyether polyols HEP-330N (molecular weight 4980, hydroxyl value 0.624mmol/g) was provided by Jining Huakai Resin Co., Ltd. 2-hydroxyethyl disulfide (HEDS, 90%), isophorone diisocyanate (IPDI, 99%) and triphenylbismuthine (TPB) were purchased from Macklin Biochemical Technology Co., Ltd, Shanghai, China. Dibutyl phthalate (DBP, 99.5%) and N, N-dimethylformamide (DMF, ⩾ 99.8%) were obtained from Damao Chemical Reagent Factory, Tianjin, China and used directly.

2.2. Preparation

In this experiment, three implementation examples and one comparative example were set. The proportion of the components in the experiment is as follows: the mass ratio of HTPB to HTPE was 4:1, the mass ratio of 330N to HEDS was 0.087, R value was 1, DBP accounts for 13% of the mass of the mixture of HTPB and HTPE, and TPB accounted for 0.1% of the mass of the mixture of HTPB and HTPE.

Implementation example 1. HTPB was mixed with HTPE to a tri-neck flask and dried in vacuo at 60℃ for 3h. Then the mixture was preheated at 70℃, and added IPDI, DBP, TPB and 3ml DMF into the flask. After stirring at 70℃ and under the protection of nitrogen for 3h, the mixture was poured into the PTFE mold and solidified at 60℃ for 36h, which was marked as a copolymerization group with the code of 2#.

Implementation example 2. The HTPB was added to a three-necked flask and vacuum dried at 60℃ for 3h. Then HTPB was preheated at 70℃, IPDI, DBP, TPB and 3ml DMF were added to the flask, and the prepolymer was obtained by stirring at 70℃ and under the protection of nitrogen for 3h. Then 330N, HEDS and 3ml DMF were evenly mixed and added into flask. After 3h vacuum drying under the protection of nitrogen at 70℃, they were poured into PTFE mold and solidified at 60℃ for 36h, which were recorded as mixed chain extender group with the code of 3#.

Implementation example 3. HTPB was mixed with HTPE to a tri-neck flask and dried in vacuo at 60℃ for 3h. Then the mixture was preheated at 70℃, IPDI, DBP, TPB and 3ml DMF were added to the flask, and the prepolymer was obtained by stirring at 70℃ and under the protection of nitrogen for 3h. Then 330N, HEDS and 3ml DMF were evenly mixed and added into flask. After 3h vacuum drying under the protection of nitrogen at 70℃, they were poured into PTFE mold and solidified at 60℃ for 36h, which was marked as the copolymerization/mixed chain extender group with the code of 4#.

Comparative example. The HTPB was added to a three-necked flask and vacuum dried at 60℃ for 3h. Then the HTPB was preheated at 70℃, and IPDI, DBP, TPB and 3ml DMF were added to the flask. After stirring at 70℃ and under the protection of nitrogen for 3h and vacuum drying, the flask was poured into the PTFE mold and solidified at 60℃ for 36h, which was recorded as pure HTPB group with the code 1#.

2.3. Characterization

According to GB/T 528-92, the tensile test was carried out at the speed of 100 mm/min on the Instron 5982 material testing machine. The dynamic characteristic analysis was carried out on DMA Q800 (TA instrument). The frequency range was 0.1Hz ~ 100Hz and the temperature was 60℃. The thermal properties of the composites were studied by differential scanning calorimeter (DSC Q20, TA instrument) in nitrogen atmosphere. The temperature range was -90 ~ 60℃ and the heating rate was 15℃/min.

3. Results and discussion

3.1. Mechanical properties

Figure 1 showed the stress-strain curve, elastic modulus, tensile strength and elongation at break of the four kinds of polyurethane. According to the stress-strain curve, it was observed that the mechanical properties of polyurethane were obviously improved by copolymerization with HTPE, adding mixed chain extender and the joint action of both. Among them, the mechanical properties of copolymer/mixed chain extender group were the best, followed by copolymerization group and mixed chain extender group. Through Figure 1(b), it can be seen that the elastic modulus of the copolymer/mixed chain extender group was the highest, up to 0.322MPa, which was 106.3% higher than that of pure HTPB (0.208MPa), while the copolymerization group and the mixed chain extender group were increased by 54.8% and 21.6%, respectively. Further, according to the calculation of the tensile strength and elongation at break of the sample given in Figure 1(c), the tensile strength of polyurethane copolymerized with HTPE was as high as 1.239MPa, which was 150.3% higher than that of pure HTPB (0.495MPa), the elastic modulus of mixed chain extender group was 1.001MPa, which was 102.2% higher than that of pure HTPB, while the elastic modulus of copolymerization/mixed chain extender group was as high as 1.263MPa, which was 155.2% higher than that of pure HTPB. On the other hand, for polyurethane, compared with 1013.15% of pure HTPB, the elongation at break of copolymerized group was 1197.88%, which was 18.2% higher than that of pure HTPB, and that of mixed chain extender group was 1268.72%, which was 25.2% higher than that of pure HTPB, while that of copolymerized/mixed chain extender group was 1585.37%, which was 56.5% higher than that of pure HTPB. It can be judged that the mechanical properties of HTPB polyurethane can be improved by the copolymerization of HTPE and HTPB and the addition of mixed chain extender 330N/HEDS, and the enhancement effect was more remarkable when both of them acted simultaneously.

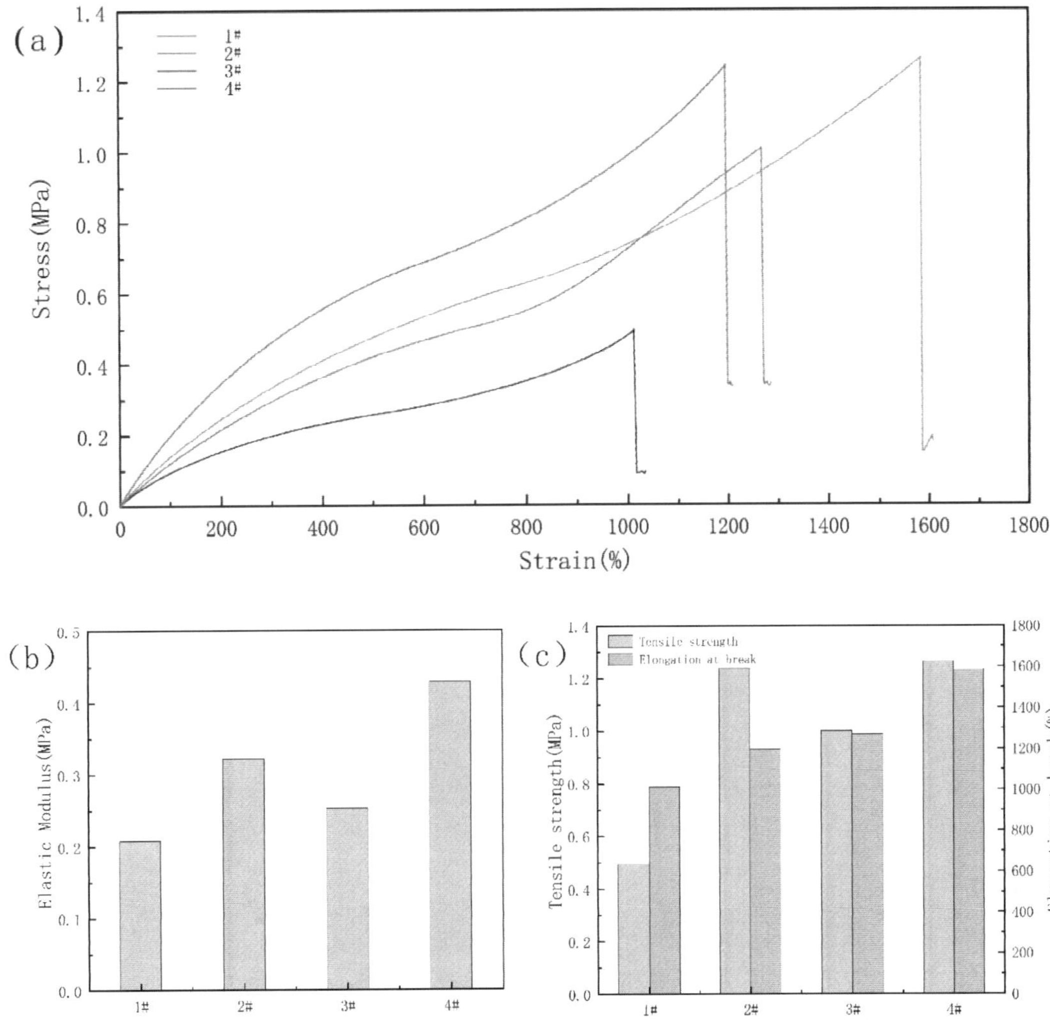

Figure 1. (a) Stress-strain curve of the polyurethane material. (b) Elastic modulus of the polyurethane material. (c) Tensile strength and elongation at break of polyurethane materials.

3.2. Dynamic mechanical properties

The dynamic mechanical properties can measure the storage modulus (G') and loss modulus (G") of the material, which provides the elastic and viscous properties of the material, and can better reflect the mechanical properties of the material [20,21]. Therefore, the dynamic mechanical analysis of polyurethane was carried out at 60℃ and frequency ranged from 0.1Hz ~ 100Hz. The storage modulus and loss modulus of polyurethane were shown in Figure 2. At 60℃, G' and G" of the four polyurethane materials all increased with the increase of frequency, and G' was greater than G", indicating that polyurethane materials were still dominated by elasticity at 60℃. In addition, it can be seen that the values of G' and G" of pure HTPB group were the smallest, and those of mixed chain extender group, copolymer group and copolymerization/mixed chain extender group increased in turn. It can be judged that copolymerization and adding mixed chain extender can improve the storage modulus and loss modulus of polyurethane, that was, they had better elastic and viscous properties, and the enhancement effect was more significant when both of them acted simultaneously.

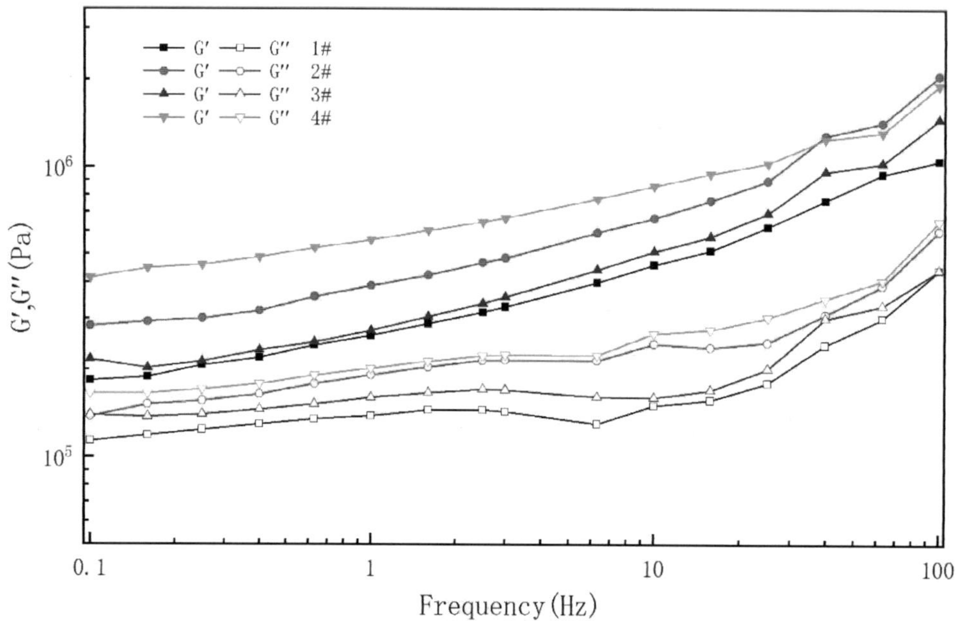

Figure 2. The relationship between storage modulus (G') and loss modulus (G") of polyurethane material and frequency at 60℃.

3.3. Thermal properties

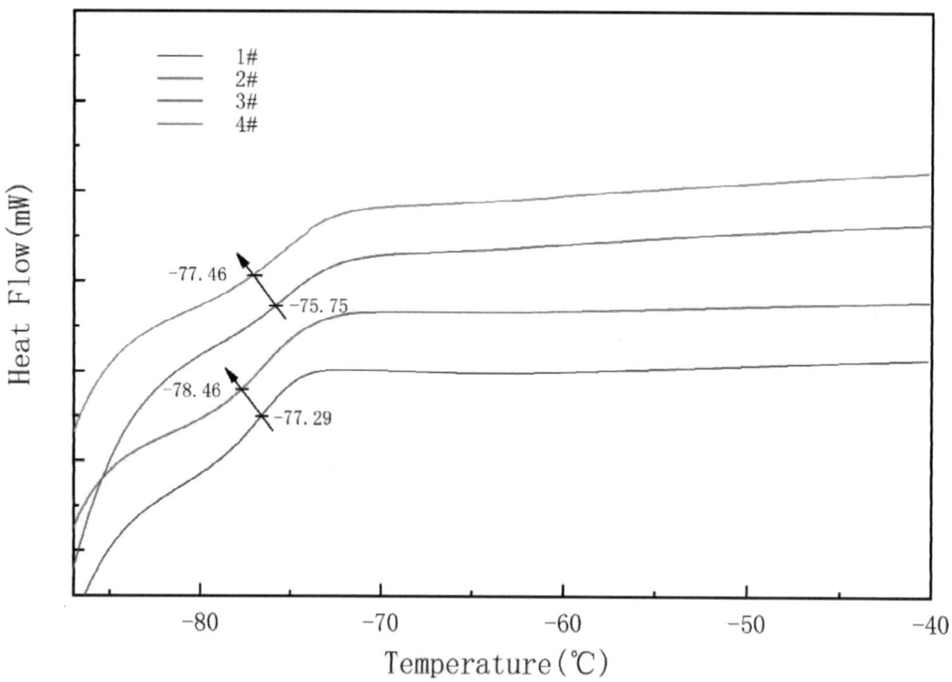

Figure 3. DSC heating curve of polyurethane material.

The glass transition temperature (Tg) is an important parameter that determines the performance of the polymer material. It is the lower limit of the use of rubber materials, which determines the temperature range of the material [22]. Figure 3 showed the DSC curve and Tg value of the polyurethane material. It can be observed from curves 1 and 2 that the glass transition temperature of HTPB polyurethane decreases from -77.29℃ to -78.46℃ by copolymerization with HTPE, which was due to the lower glass transition temperature of HTPE itself, which was about -80℃. Secondly, the crystallization of polymer can be reduced and the flexibility of molecular chain can be enhanced by copolymerization, thus reducing the glass transition temperature. It can be seen from the curve 1, 3 that the Tg value of HTPB polyurethane was increased by adding a mixing chain extender, from -77.29℃ to -75.75℃, which was because polyether polyols HEP-330N was a macromolecular chain extender, molecular weight very large, and the molecular chain contained an ether bond, but it contained three hydroxyl groups, which form crossover points, block the movement of the molecular chain, thereby rose the glass transition temperature [18]. According to curves 1 and 4, when the copolymerization and the mixed chain extender acted together, the glass transition temperature of polyurethane reached a compromise and basically remained unchanged, ranging from -77.29℃ to -77.46℃. It can be seen that the Tg values of polyurethane were all in the range of -79℃ ~ -75℃, which was far lower than room temperature, ensuring the excellent low temperature mechanical properties of polyurethane, which was of great significance for its application.

4. Conclusion

In summary, this paper studied the effects of chemical copolymerization and mixed chain extenders on HTPB polyurethaneological properties. The stress-strain curve, elastic modulus, tensile strength and elongation at break of modified polyurethane were obtained by static tensile test. The results showed that both the copolymerization and the addition of mixed chain extender can improve the mechanical properties of the polyurethane, and the improvement effect was more remarkable when both were applied simultaneously. DMA was used to test the dynamic mechanical properties of polyurethane, and the relationship between G', G" and frequency of polyurethane at 60℃ was obtained. The results showed that copolymerization and adding mixed chain extender can improve the storage modulus and loss modulus of polyurethane, that was, they had better elastic and viscous properties, and the enhancement effect was more significant when both of them acted simultaneously. The glass transition temperature of the polyurethane was measured by differential scanning calorimetry, and the results showed that Tg was substantially unchanged, all of which were in the range of -79℃ to -75℃, which were much lower than room temperature to ensure excellent low temperature mechanical properties of polyurethane. This work provided an effective method and strategy for preparing high performance polyurethane materials.

References

[1] Xiaochuan Wang YS, Yonglin Lei, Jichuan Huo, Xianming Lu, Qian Zhang, Yingying Lu and Wei Wang. Trimethylisophthtalodinitrile Oxide as a Room-Temperature Curing Agent for Curing HTPB. *Advanced Functional Materials*. 2018.

[2] Ke Chen QR, Jinjin Li, Dafa Chen, Chunxiang Li. A highly stretchable and self-healing hydroxyterminated polybutadiene elastomer. *Journal of Saudi Chemical Society*. 2020.

[3] Pingan Zhang WT, Xilong Zhang, Juan Chen, Jianmin Yuan, and Jianru Deng. Chemical Modification of Hydroxyl-Terminated Polybutadiene and Its Application in Composite Propellants. *Industrial&Engineering Chemistry Research*. 2021.

[4] Lu Dai SJ, Zhiyang Bu, Bo-Geng Li. Supramolecular thermoplastic elastomers via selfcomplementary quadruple hydrogen bonding between polybutadiene-based triblock copolymers. *Journal of Applied Polymer Science*. 2020.

[5] Ali Nazmi MY, Mona Zamani Pedram, Mansour Shahidzadeh. Evaluation of the 3-hydroxy pyridine antioxidant effect on the thermal-oxidative degradation of HTPB. *International Journal of Chemical Kinetics*. 2018.

[6] Qinzhuo Zhou SJ, Bo-Geng Li. Facile synthesis of novel HTPBs and EHTPBs with high cis-1,4 content and extremely low glass transition temperature. *Polymer*. 2015.
[7] Yi-Song He J-BZ, Guang-Chen Liu, Qiu-Tong Li and Yu-Zhong Wang. Super-tough poly(L-lactide)/crosslinked polyurethane blends with tunable impact toughness. *RSC Advances*. 2014.
[8] F. Shaghayeghi Toosi MS, B. Ramezanzadeh An investigation of the effects of pre-polymer functionality on the curing behavior and mechanical properties of HTPB-based polyurethane. *Journal of Industrial and Engineering Chemistry*. 2015.
[9] Qin Ning, Liu Yucun, Chai Tao, Guo Jiaqu, Ma Hui, Li Jinlong, Wushu. Preparation and properties of hydroxyl-terminated polybutadiene (HTPB)-polyurethane elastomer. *Insulating materials*. 2016. (In Chinese)
[10] Gao Yuxin, an Yanjie, Jiang Yanfeng, Li Rui. Preparation and characterization of propylene-1-hexene random copolymer. *Synthetic resins and plastics*. 2021. (In Chinese)
[11] Yang Xiuling, Zhang Zhining, Zhang Delong, Li Haoyang, Yuan Shicai, Wang Xiulei, Yang Song. *Improvement of mechanical properties of PVC resin*. 2020. (In Chinese)
[12] Keke Chen XW, Guoping Li, Siping Pang and Yunjun Luo. Improvement of mechanical properties of in situprepared HTPE binder in propellants. *RSC Advances*. 2020.
[13] Keke Chen SY, XiaoMu Wen, Chao Sang and Yunjun Luo. Effect of Mixed Isocyanate Curing Agents on the Performance of In Situ-Prepared HTPE Binder Applied in Propellant. *Propellants Explosives Pyrotechnics*. 2020.
[14] Yang Chongchong. Study on properties and application of HTPE modified GAP adhesive system. 2016. (In Chinese)
[15] Ul-Haq NAMBKXMN. The Influence of Cross-Linking/Chain Extension Structures on Mechanical Properties of HTPB-Based Polyurethane Elastomers. *Arabian Journal for Science and Engineering*. 2014, **39**(1): 43-51.
[16] Liu Xiaowen, Hua Xueqing, Liu Jinchun. Factors affecting the glass transition temperature of polyether polyurethane elastomers. *Thermosetting resin*. 2018. (In Chinese)
[17] Liu Xiaowen, Hua Xueqing, Qin Xianyu, Liu Jinchun. Study on mechanical properties of polyether polyurethane elastomer of PTMG system. *Special rubber products*. 2018. (In Chinese)
[18] Fu Huijuan, Yu Chengke, Luo Jianxun, Mao Lixin, Zhang Liqun. Effect of chain extender and cross-linking agent on properties of NDI polyurethane elastomer. *China Plastics*. 2012. (In Chinese)
[19] Luo Jianxun, Jin Hao, Mao Lixin, Zhang Liqun. Effect of mixed chain extender on macroscopic properties of polyurethane elastomers. *Journal of Beijing University of Chemical Technology* (Natural Science Edition). 2011. (In Chinese)
[20] Lakshmana Rao Bhagavathi APD, G.D. Janaki Ram, S.K. Panigrahi. Hygrothermal aging, fatigue and dynamic mechanical behavior of cellulosic particles reinforced one-component moisture curable polyurethane adhesive joints. *International Journal of Adhesion and Adhesives*. 2021.
[21] Tayebe Khanlari YB, Mohammad Bayat, Nasser Sheibani Synthesis, characterization, and curing of propylene oxide and glycidyl nitrate random copolymer (GN-ran-PO) and investigation of its compatibility with different energetic plasticizers. *Journal of Molecular Structure*. 2021.
[22] Sun Zongxue, Zhang Xiaolian, Zhou Zhifeng, Ji Meiqin, Zhang Huan, Chen Hong, Cai Qing. Research progress of cold-resistant rubber materials. *Rubber industry*. 2021. (In Chinese)

IOP Conf. Series: Materials Science and Engineering 1167 (2021) 012013 doi:10.1088/1757-899X/1167/1/012013

Effect of water- and acid-pickling of aquaticplant biochar on the adsorption of Cd

Wenbing Tan

State Environmental Protection Key Laboratory of Simulation and Control of Groundwater Pollution, Chinese Research Academy of Environmental Sciences, Beijing 100012, China

Corresponding Author's E-mail: tanwenbing@pku.edu.cn

Abstract. To reveal the influencing factors and mechanism of adsorption of heavy metal Cd by biochar, different elution solvents (water and dilute acid solution) were used in this paper to elute the residual ash substances after pyrolysis of biochar, then the pH was regulated to investigate the adsorption effect of biochar on Cd in different treatment conditions. The precipitation experiments of Cd by biochar soaking and washing filtrate showed that the precipitation of root, stem and leaf of *Acorus calamus L.* contributed a great deal to the heavy metal removal system, which accounting for 81%, 77% and 98%, respectively, and the adsorption effect of biochar of roots, stems and leaves of *Acorus calamus L.* on Cd was slightly increased after washing. The adsorption experiment of biochar pickling on heavy metals showed that the adsorption effect of biochar on heavy metals Cd decreased greatly after pickling, which may be due to the removal of most of the minerals on the surface of biochar after pickling and the decrease of pH value of biochar. This work can provide insights into the use of biochar to control Cd.

1. Introduction

During the preparation of biochar, ash, inorganic salts, tar, vinegar and gaseous methane are also produced along with the solid carbon. In addition to large amounts of cellulose and lignin, biomass is also rich in a variety of trace metal elements, such as K, Ca, Na, Mg, Si, Fe, B and others, and these metal elements will be condensed into ash in the pyrolysis process and then remain in the carbon. The different pyrolysis conditions and precursors will affect the different contents of oxides and alkali metal salts in ash. For example, ash is inert in gas phase adsorption and has no effect on the adsorption process. However, in liquid adsorption medium, the content of oxides and alkali metal salts in ash will affect the adsorption process. In addition, the adsorption of heavy metal ions by biochar after ash removal was significantly reduced, indicating that ash in biochar plays an important role in the removal of environment heavy metal ions [1, 2].

It has been reported that ash affects the adsorption of heavy metals by biochar mainly through the following methods: (1) After soluble in water, the ash components of biochar can release a certain amount of PO_4^{3-}, CO_3^{2-} and SO_4^{2-}, which can combine with heavy metal ions to form precipitates [3, 4]. (2) SiO_2 particles in ash can adsorb heavy metal ions [5]. (3) High ash biochar is highly alkaline, which can make heavy metal ions in the aqueous solution form ammonia oxide precipitation, thus reducing the concentration of heavy metal ions in the aqueous solution [6]. (4) When the biochar adsorbed heavy metal ions, the contents of K^+, Na^+, Ca^{2+} and Mg^{2+} increased compared with the system before the adsorption, indicating that the basic cations (K^+, Na^+, Ca^{2+} and Mg^{2+}) in the biochar ash components could exchange cations with heavy metal ions in the aqueous solution [7, 8]. (5) The soluble cation contained in biochar will compete with heavy metals for adsorption sites on biochar, thus reducing the adsorption capacity of biochar on heavy metals. Therefore, the influence of ash content and mineral composition in biochar on adsorption need be studied.

In order to study the influence of biochar on the adsorption of heavy metal Cd and its mechanism, different elution solvents (water and dilute acid solution) were used in this paper to elute the residual ash substances after pyrolysis of biochar, and then the adsorption effect of biochar on Cd was compared in different treatment conditions through adjusting pH.

2. Materials and methods

2.1. pH and precipitation of heavy metals in biochar aqueous solution

The biochar was mixed with pure water at the ratio of 1 g/L, and then oscillated in a constant temperature oscillation chamber for 2 h. The biochar was separated from the cleaned filtrate through a suction filter, then the cleaning filtrate of root, stem and leaf of *Acorus calamus L.* is obtained. The pH of the supernatant was measured with a pH meter, and the initial pH in root, stem and leaf of *Acorus calamus L.* was recorded. The alkalinity of biochar lotions was determined by acid-base titration, and the alkalinity of root, stem, and leaf was recorded. In addition, several sets of 20 mL cleaning filtrate of root, stem and leaf of *Acorus calamus L.* were respectively taken and placed in 20 mL of Cd stock solution with a concentration of 20 mg/L $Cd(NO_3)_2 \cdot 4H_2O$, and then oscillated in a thermostatic oscillation chamber at 25 °C. The content of Cd in heavy metals in the solution was measured after a 0.22 μm microporous filtration membrane. Finally, the adsorption amount of Cd was calculated by the following equation:

$$Q_e = \frac{(C_0 - C_e)V}{m} \quad (1)$$

2.2. Adsorption of heavy metals on biochar after washing and drying

The biochar was mixed with pure water at the ratio of 1 g/L, and the biochar was oscillated in a constant temperature oscillation chamber at 25 °C for 2 h. The biochar was separated from the cleaned filtrate with a suction filter device. The biochar was placed on a filter paper and dried in an oven at 70 °C for 24 h, so as to obtain the biochar content in root, stem and leaf of *Acorus calamus L.* after washing and drying. Then 0.02 g of washed and dried biochar was placed in 20 mL Cd stock solution with a concentration of 20 mg/L $Cd(NO_3)_2 \cdot 4H_2O$. The biochar was oscillated in a thermostatic oscillation chamber at 25 °C and filtered through a 0.22 μm microporous membrane. Finally, the content of Cd in the solution was measured. The adsorbing capacity of Cd was calculated according to the equation (1).

2.3. Adsorption of heavy metals on biochar after pickling and drying

1 g of biochar powder was weighed and placed in a beaker, and then stirred flushed with 1 mol/L HCl solution, repeated three times, the total amount of acid was 1 L, and the reaction time of each decarboxylation was 30 min, then rinse with deionized water for several times, until the pH of supernatant is consistent with that of pure water pH (pH = 5.3), then the sample was dried overnight at 70 °C, and the biochar content in root, stem and leaf of *Acorus calamus L.* after pickling and drying was obtained. Furthermore, 0.02 g of pickled and dried biochar was placed in 20 mL of $Cd(NO_3)_2 \cdot 4H_2O$ stock solution with Cd concentration of 20 mg/L, the solution was placed in a constant temperature oscillation box to oscillate at 25 °C and filtered through 0.22 μm microporous membrane. Finally, the content of Cd in the solution was measured. The adsorption capacity of Cd was calculated according to the equation (1).

2.4. The adsorption of heavy metals on the original pH after biochar pickling and drying

0.02 g of pickled and dried biochar in root, stem and leaf of *Acorus calamus L.* was put into 20 mL

$Cd(NO_3)_2{\cdot}4H_2O$ stock solution with the concentration of 20 mg/L $Cd(NO_3)_2{\cdot}4H_2O$. The solution was adjusted to the same initial pH with 0.01 mol/L NaOH. Then, the solution was oscillated in a constant temperature oscillation chamber at 25 °C and filtered through a 0.22 μm microporous membrane. Finally, the amount of Cd in the solution was measured. The adsorption amount of Cd was calculated according to the equation (1). The plasma concentrations of Cd^{2+}, Ca^{2+}, Mg^{2+} and K^+ in the solution were measured by inductively coupled plasma mass spectrometry.

3. Results and discussion

3.1. pH of biochar water lotion from Acorus calamus L.

The initial pH of root, stem and leaf of *Acorus calamus L.* is 9.5, 10.2 and 10.1, respectively. The pH of biochar from stem and leaf of *Acorus calamus L.* was similar, which was higher than that from root. This was positively correlated with the amount of Cd adsorbed by biochar from root, stem and leaf of *Acorus calamus L.* in different initial concentrations. In the case of low pH value, H^+ had a strong competitive effect on Cd, on the other hand, when the acidity is strong, the surface functional groups of biochar are protonated, leading to the occupation of the adsorption site [9–11]. When the alkalinity of the solution increases, the protonation of functional groups decreases, and the electrostatic repulsion between biomass and Cd also decreases. The adsorption is partly due to the partial precipitation of the solution near the surface of biochar, and partly due to the cation exchange between Cd and biochar matrix alkaline earth metal ions.

3.2. Influence of water pickling of biochar on heavy metal precipitation

In the process of biomass pyrolysis, potassium oxide, calcium oxide and magnesium oxide are formed. These oxides react with water then hydroxides are produced. The plasma of K^+, Na^+, Ca^{2+} and Mg^{2+} will be released from the biochar, so biochar is always alkaline. The release of these cations will compete for the adsorption sites of Cd on biochar. The water-soluble substances in ash will be removed when wash with water, and washing with acid can remove the insoluble substances dissolved in acid and reduce the ash content to a greater extent. The alkalinity of biochar will be reduced through washing or pickling, the ash content and pH of biochar will change greater by pickling than by washing. The alkaline earth metals (Ca and Mg) of biochar lotion from root of *Acorus calamus L.* is 2.8 mg/L, and the content of K is 3.6 mg/L. And the alkaline earth metals (Ca and Mg) of biochar lotion from stem of *Acorus calamus L.* is 4.0 mg/L, and the K content is very prominent (85.2 mg/L). Furthermore, the alkaline earth metals (Ca and Mg) of biochar lotion from leaf of *Acorus calamus L.* is 10.3 mg/L, and the content of K is 66.6 mg/L. This is due to the fact that the content of alkaline earth metals (Ca and Mg) in stem and leaf of *Acorus calamus L.* is relatively high, the solubility of the metal enhanced the alkalinity of the solution, and the target pollutant Cd is then precipitated and removed.

In order to evaluate the contribution of biochar to the precipitation of heavy metals in the removal of heavy metals system, the biochar lotion was used to precipitate Cd. As shown in Figure 1, precipitation contributed greatly to the heavy metal removal system, accounting for 81%, 77% and 98%, respectively. The results showed that the adsorption of Cd by leaf of *Acorus calamus L.* was higher than that by the stem under various treatment conditions, while the adsorption by the root of *Acorus calamus L.* was the least. For the biochar in root, stem and leaf of *Acorus calamus L.*, the adsorption capacity of biochar on Cd increased after washing, and the adsorption capacity of biochar on Cd increased obviously after washing. This indicated that the soluble ions are removed after washing, reducing the competitive effect of cations (K^+, Ca^{2+}, Mg^{2+}) on Cd in the adsorption, thus slightly increasing the adsorption capacity of biochar on Cd. This implied that in the process of heavy metal adsorption of biochar in root of *Acorus calamus L.*, the biochar after washing and drying has the same adsorption performance for Cd as the biochar after pickling and adjusting the solution pH to the original pH. Indirectly, the soluble ions were washed away after washing, reducing the competition for Cd, leading to a slight increase in Cd. When the pH raised, the target pollutant Cd can be removed by precipitation, and the effect of biochar on Cd is dominated by precipitation. The adsorption of Cd by biochar after pickling was significantly decreased, and the adsorption of Cd in root, stem and leaf of *Acorus calamus L.* by folic acid washing was decreased by 72%, 75% and 64%, respectively. After pickling, inorganic minerals insoluble in water on the surface of biochar were removed, indicating that inorganic minerals of biochar played an important role in Cd, which was consistent with the research results. H+ reacts with anions such as HCO_3^-, CO_3^{2-} and SO_4^{2-} in biochar, which reduces the alkalinity of biochar solution and the precipitation with Cd, thus reducing the adsorption of Cd [12, 13]. The adsorption capacity of Cd in leaf of *Acorus calamus L.* was significantly improved when the pH was adjusted to the original pH after washing and pickling.

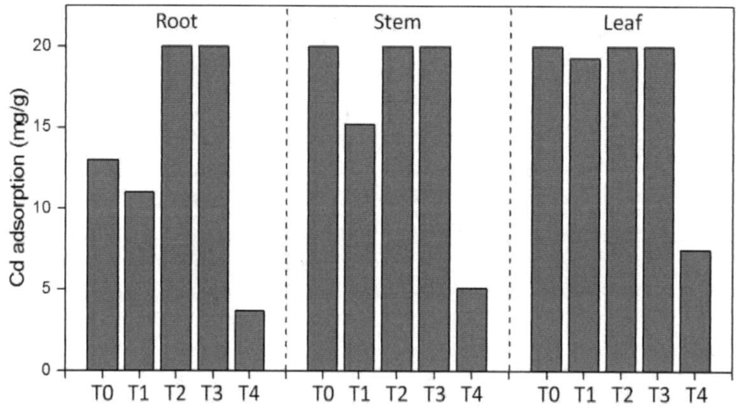

Figure 1. Influence of biochar on cadmium adsorption in different elution treatment conditions. T0 indicates the untreated biochar, T1 indicates the biochar lotion, T2 indicates the biochar after washing and drying, T3 indicates the pH adjusted to the original pH after pickling, T4 indicates the biochar after pickling.

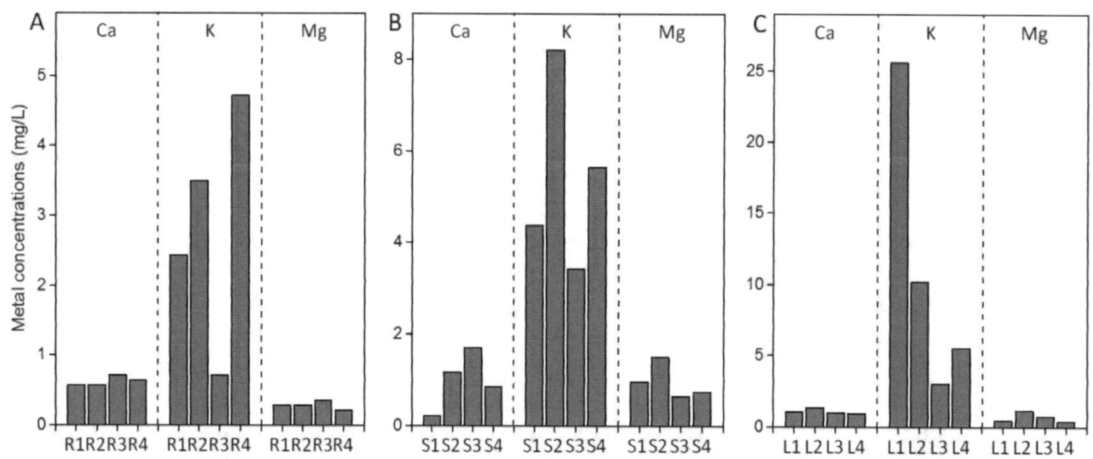

Figure 2. Changes in concentrations of common metals in the solution after adsorption of Cd by biochar of *Acorus calamus L.* in different elution conditions. A, B and C indicate the biochar ion concentration variation in root, stem and leaf of *Acorus calamus L.*, respectively. R1, S1 and L1 were untreated biochar in root, stem and leaf of *Acorus calamus L.*, respectively. R2, S2 and L2 were biochar in root, stem and leaf of *Acorus calamus L.* after washing and drying, respectively. R3, S3 and L3 were pickled and dried biochar in root, stem and leaf of *Acorus calamus L.*, respectively. R4, S4 and L4 were pickled and adjusted to the original pH biochar in root, stem and leaf of *Acorus calamus L.*, respectively.

As shown in Figure 2, Ca^{2+}, K^+ and Mg^{2+} all increased after the adsorption of Cd by the biochar of *Acorus calamus L.* in different elution conditions, indicating that Cd and base ions on the surface of biochar were ion exchanged, normally alkali or alkaline earth metals, resulting in the increase of base ions in the solution and the removal of Cd. After pickling, the surface minerals of biochar were removed, resulting in the decrease in the adsorption capacity of Cd and the base ions in the solution were less than those in other treatment conditions. It was found that K^+ precipitated most in the solution when the biochar adsorbed Cd in different treatment conditions, indicating that K^+ on the three kinds of biochar was most likely to exchange with Cd. After washing, the adsorption of Cd on biochar and the increase of various ions in the solution was more obvious, which further illustrated that the soluble ions were washed away by washing, and the adsorption competition with Cd was reduced.

Plant wastes mainly include straw, straw, and branches, which are usually rich in carbon. If these wastes are burned directly, a large amount of carbon emissions could be generated, which will not only cause waste of resources, but also damage the ecological environment. Thus, they can be made into biochar for further utilization. Our results indicate that the adsorption performance of biochar can be greatly improved through reasonable pretreatment of aquaticplant biochar, which can significantly promote the treatment of heavy metal pollution in the environment.

4. Conclusions

In this paper, different elution solvents (water and dilute acid solution) were used to study the residual ash substances after the pyrolysis of eluted biochar. The results showed that biochar had a significant effect on the adsorption process of Cd, the conclusions were as follows:

(1) According to the removal of Cd by the water-washed filtrate of biochar, it was found that in addition to the role of its own structure and surface functional groups on biochar, the water-soluble parts of the biochar from root, stem and leaf of *Acorus calamus L.*, such as HCO_3^-, CO_3^{2-}, and the precipitation with OH^- in the solution contributed 81%, 77% and 98% to the removal of Cd, respectively.

(2) According to the removal of Cd by the water-washed filtrate of biochar, it is found that the adsorption effect of the biochar in root, stem and leaf of *Acorus calamus L.* on Cd increased slightly after water washing. It indicated that on the one hand, some soluble ions were washed away after water washing, and the adsorption competition for Cd was reduced. On the other hand, it provided more sites for the adhesion of Cd by dredging the blocked pores on the biochar surface.

(3) According to the removal of Cd by biochar after pickling, it was found that the adsorption effect of biochar on Cd was greatly reduced after pickling, most of the surface minerals of biochar were removed, and the pH of biochar was also reduced, which resulted in a decrease in the adsorption capacity of Cd.

Acknowledgments

The author thanks Tongtong Li and Lei Wang for assistance in the laboratory and with data analysis. This work was financially supported by the National Natural Science Foundation of China (51908375).

References

[1] Wang Z, Liu G, Zheng H, Li F, Ngo HH, Guo W, Liu C, Chen L and Xing B 2015 *Bioresour. Technol.* **177** 308–317.
[2] Munir MAM, Liu G, Yousaf B, Ali MU, Abbas Q and Ullah H 2020 *Chemosphere* **240** 124845.
[3] Xu X, Cao X and Zhao L 2013 *Chemosphere* **92** 955–961.
[4] Yang F, Sui L, Tang C, Li J, Cheng K and Xue Q 2021 *Sci. Total Environ.* **768** 145106
[5] Yang H, Xu R, Xue X, Li F and Li G 2008 *J. Hazard. Mater.* **152** 690–698.
[6] Jiang J, Peng Y, Yuan M, Hong Z, Wang D and Xu R 2015 *Pedosphere* **25** 781–789.
[7] Jiang S, Huang L, Nguyen TAH, Ok YS, Rudolph V, Yang H and Zhang D 2016 *Chemosphere* **142** 64–71.

[8] Feng D, Zhang Y, Zhao Y and Sun S 2018 *Energy* **152** 166–177
[9] Ofomaja AE, Unuabonah EI and Oladoja NA 2010 *Bioresour. Technol.* **101** 3844–3852.
[10] Chen Z, Xiao X, Chen B and Zhu L 2015 *Environ. Sci. Technol.* **49** 309–317.
[11] Qian L, Zhang W, Yan J, Han L, Gao W, Liu R and Chen M 2016 *Bioresour. Technol.* **206** 217–224.
[12] Boyd SA and Shelton DR 1984 *Appl. Environ. Microbiol.* **47** 272–277.
[13] Kappler A, Wuestner ML, Ruecker A, Harter J, Halama M and Behrens S 2014 *Environ. Sci. Technol.* **1** 339–344.

IOP Conf. Series: Materials Science and Engineering 1167 (2021) 012014 doi:10.1088/1757-899X/1167/1/012014

Effect of softening in near-surface region on wear transition of Ti-6Al-4V alloy at elevated temperatures

D H Du[1], Z J Yu[1], F J Tao[1] and J An[1,2]

[1]School of Materials Science and Engineering, Jilin University, Changchun 130025, PR. China

[2]Corresponding Author's E-mail: anjian@jlu.edu.cn

Abstract. Dry sliding wear testing of Ti-6Al-4V alloy was performed at temperatures of 20, 100 and 200°C to investigate it's wear transition mechanism. Each wear rate-applied load curve demonstrated a mild wear stage and a severe wear stage. Morphology and composition of worn surface were analyzed using SEM and EDS techniques. When wear transformed from the mild wear stage to severe wear stage, sever plastic deformation wear mechanism played a dominant role. The analysis of subsurface microstructures revealed that a fine grain microstructure zone was formed beneath the surface after mild to severe wear transition, correspondingly a soften effect was detected in the fine grain microstructure zone by microhardness measurement. Therefore, it is considered that the softening in the near surface is the main reason for mild to severe wear transition.

1. Introduction

Because of the excellent mechanical properties, perfect corrosion resistance and low density, titanium alloys have found wide application in biomedical implants, aviation and aerospace industries [1-4]. Ti-6Al-4V alloy is the first generation of high-temperature titanium alloy. Since its successful development in 1954, Ti-6Al-4V alloy has taken over 50% of titanium alloys consumption [5]. Nevertheless, their poor wear performance has become a major obstacle to finding applications in tribological fields. In spite that a variety of surface modification technologies have been applied for improvement of wear properties of titanium alloys, tribologists are still concerned about the reason that hides behind the poor wear performance.

A lot of efforts have been made on investigation of wear behavior of titanium alloys. Tribo-layers are an attractive subject among numerous research works. Tribo-layers form on the surface owing to the interaction between friction pairs and environment. On one hand, tribo-layers separate the titanium alloys from counterface, and prevent the direct contact between friction pairs. On the other hand, due to their high hardness, tribo-layers can maintain a low wear rate until their large-scale spallation, i.e. mild wear to severe wear transition. The positive influence of tribo-layers has been found in titanium alloys under low and high speeds or elevated temperatures of 300°C and 400°C [6-8]. However, under intermediate speeds or elevated temperatures of 50-200°C, tribo-layers are usually absent on the surfaces, in this situation the mechanism for rapid rising of wear rate or mild to severe wear transition is still not clarified.

The present study is aimed at investigate the wear behavior and mechanism for mild to severe wear transition of Ti-6Al-4V alloy. The effect of softening in the near-surface region during sliding wear was mainly discussed.

2. Experimental details

The studied material was a hot-forged Ti-6Al-4V alloy with α-Ti and β-Ti phase. Its microstructure was comprised of equiaxed grains of α-Ti phase and continuous network of β-Ti phase. The alloy had a Vikes hardness of 310HV.

Wear tests were performed on a pin-on-disk wear tester at 1.0 m/s under unlubricated conditions. Pin samples were machined form as-received Ti-6Al-4V alloy into dimensions of 6 mm in diameter and 13 mm in length. Disks were made form 50Cr steel into dimensions of 70 mm in diameter and 13 mm in thickness with a harness of HRC56. Test temperatures were 20°C, 100°C and 200°C, applied load ranged from 10 N to 180 N, sliding distance was 565 m. Hardness change in the surface and subsurface were also analyzed using microhardness tester, respectively.

3. Results and Discussion

3.1. Wear rates

The wear rates of Ti-6Al-4V alloy were shown in figure 1 as a function of applied load under different test temperatures. Under each test temperature, wear rate experienced two stages with increasing load, i.e. a fast-and-gradual increasing stage and a rapid rising stage. Additional, test temperature exhibited a strong influence on the wear rate, namely raising temperature lead to an increase of wear rate. Apparently, the first stage corresponded to the mild wear, while the second stage corresponded to the severe wear stage. With increasing test temperature from 20°C to 200°C, the critical load for mild to severe wear transition decreased from 105 N to 70 N.

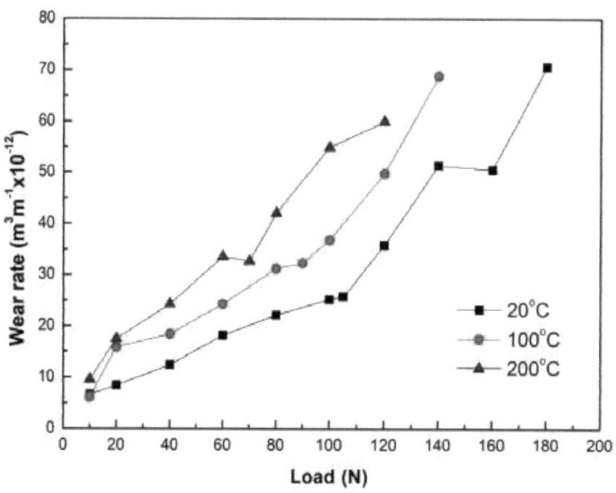

Figure 1. Wear rates vs. applied load under different test temperatures.

3.2. Wear mechanisms and surface hardness

The wear mechanisms of Ti-6Al-4V alloy were identified by means of SEM and EDS techniques. The main wear mechanisms were oxidation, spallation of oxide layer, slight plastic deformation and adhesion in the mild were stage, and severe plastic deformation and adhesion in the severe wear stage. Figure 2 shows the worn surface morphologies in the mild and severe wear stages. As test temperature was 100°C, at 80 N in the mild wear stage, the surface was slightly deformed with a flat appearance (figure 2(a)), and the surface was not heavily oxidzed since the content of oxygen element was 3.51%. As the load was increased to 120 N in the severe wear stage, the surface was severely deformed, which led to a flatten surface together with irregular adhesion scars (figure 2(b)). Massive surface material was plastically deformed along the sliding direction, and an extruded edge was found to be formed as shown in the magnification image in the top right corner of figure 2(b). Meanwhile, the content of oxygen element still kept as 4.54% on the surface, suggesting that surface oxidation cannot prevent mild to severe wear transition from happening. As test temperature was 200°C, at 40 N, the surface also experienced slight plastic deformation, meanwhile the adhesion scars took over almost half of the surface (figure 2(c)). The content of oxygen element was 3.23%, suggest a moderate surface oxidation. As the load was increased to 80 N in the severe wear stage, the surface was severely deformed, and the adhesion was also serious (figure 2(d)). In addition, an extruded edge can be observed in the magnification image in the top right corner of figure 2(d). As revealed by the analysis of wear mechanisms in the mild and severe wear stages, the mild to severe wear transition was controlled by the wear mechanism of severe plastic deformation.

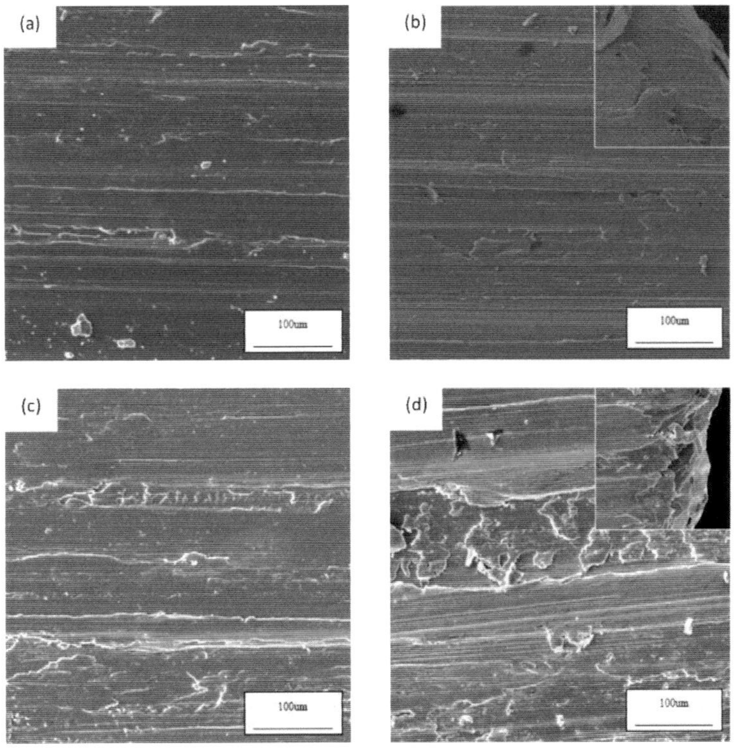

Figure 2. SEM images of worn surface at different sliding conditions: (a) 100°C and 80 N, (b) 100°C and 120 N, (c) 200°C and 40 N, (d) 200°C and 80 N.

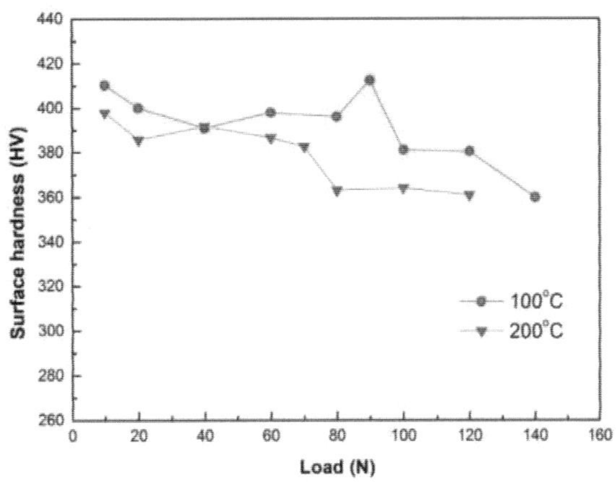

Figure 3. Worn surface hardness vs. applied load.

According to tribological theory, the surface harness of metallic materials has a great influence on wear resistance. For the purpose of comparing the difference of surface hardness between mild and severe wear stages, surface hardness of pins worn at 100°C and 200°C was measured and plotted as a function of applied in figure 3. As compared with the original hardness of Ti-6Al-4V alloy, the hardness of worn surface was increased with different degrees owing to plastic deformation and presence of oxide layer. An amazing point is that the hardness maintains a high level above 380 HV in the mild wear stage, whereas it begins to decrease to a low level of about 360 HV in the severe wear stage, suggesting the softening of surface material. Apparently, the softening of surface material is not merely caused by absence of oxide layer, since the contents of oxygen element are 5.77% at 100 N and 100°C, and 3.75% at 80 N and 200°C, respectively. The possible reason for the softening of surface material in the severe wear stage could be a type of microstructure transformation.

3.3. Subsurface microstructure and hardness

Figure 4 shows the subsurface microstructures before and after mild to severe wear transition at 100°C and 200°C. When test temperature was 100°C, at 60 N in the mild wear stage, α-Ti grains in the near-surface region were plastically deformed towards surface, forming a plastic deformation zone of about 95μm (figure 4(a)). At 120 N in the severe wear stage, an extraordinary microstructure zone with a thickness of about 18 μm occurred underneath the surface, in which an amount of fine grains were mixed with deformed grains (figure 4(b)). Underneath the fine-grain microstructure zone, there was a plastic deformation zone. When test temperature was 200°C, the change of microstructure in the near-surface region was similar to those at 100°C. At 40 N in the mild wear stage, a plastic deformation zone was formed beneath the surface (figure 4(c)), whereas at 80 N in the severe wear stage, a fine-grain microstructure zone of about 11μm thickness was also observed on a plastic deformation zone (figure 4(d)). The mechanism for the fine grain microstructure transformation needs further investigation. However, the microstructure transformation form deformed grain to fine grain could be the reason for the surface softening in figure 3.

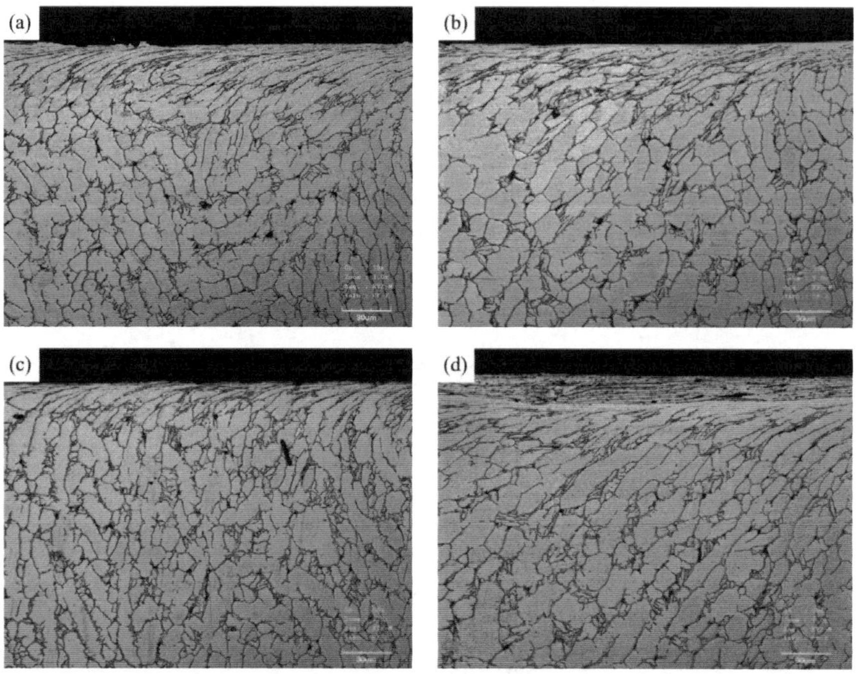

Figure 4. Subsurface microstructures at different sliding conditions: (a) 100°C and 60 N, (b) 100°C and 120 N, (c) 200°C and 40 N, (d) 200°C and 80 N.

To verify this, the microhardness was also measured in the subsurfaces a function of depth, as shown in figure 5. It can be seen that under the test temperature of 100°C, the hardness is lower at 120 N than at 60 N within the depth range of 5-20μm, similarly, under 200°C, the harness is also lower at 40 N than at 80 N within the depth range of 5-10μm. These results prove that there exists a softening effect in the near-surface region in the severe wear stage, which is gives rise to the mild to severe wear transition.

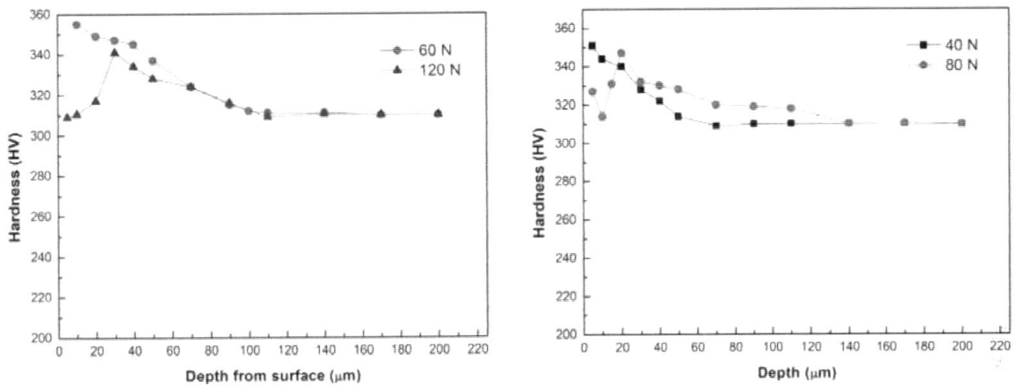

Figure 5. Hardness change in subsurfaces at different temperatures: (a) 100°C, (b) 200°C

4. Conclusions

1. Under given sliding conditions, wear rate of Ti-6Al-4V alloy increased with increasing applied load or test temperature.
2. The mild to severe wear transition was controlled by the wear mechanism of severe plastic deformation. Transition load decreases with increasing temperature.
3. The occurrence of softening in the near-surface region is the reason for mild to severe wear transition.

Acknowledgments
The authors thank the support from National Science Foundation of China (Grant No.51775226).

References

[1] Magaziner R S, Jain V K and Mall S 2008 Wear characterization of Ti–6Al–4V under fretting–reciprocating sliding conditions *Wear* **264** 1002–14
[2] Ohidul A and Haseeb A S M A 2002 Response of Ti–6Al–4V and Ti–24Al–11Nb alloys to dry sliding wear against hardened steel *Tribol. Int.* **35** 357–62
[3] Jin H X, Wei K X, Li J M, Zhou J Y and Peng W J 2015 Research development of titanium alloy in aerospace industry *Chin. J. Nonferrous. Met.* **25** 280–92 (In Chinese)
[4] Lindvall R, Lenrick F, Persson H, M'saoubi R, Ståhl JE and Bushiya V 2020 Performance and wear mechanisms of PCD and pcBN cutting tools during machining titanium alloy Ti6Al4V *Wear* **203329** 454-455
[5] Cui C, Hu B M, Zhao L and Liu S 2011 Titanium alloy production technology, market prospects and industry development *Mater. Des.* **32** 1684–91
[6] Farokhzadeh K and Edrisy A 2016 Transition between mild and severe wear in titanium alloys *Tribol. Int.* **94** 96–111
[7] Wang L, Zhang Q Y, Li X X, Cui X H and Wang S Q 2014 Severe-to-mild wear transition of titanium alloys as a function of temperature *Tribol. Lett.* **53** 511–20
[8] Mao Y S, Wang L, Chen K M, Wang S Q and Cui X H 2013 Tribo-layer and its role in dry sliding wear of Ti-6al-4V alloy *Wear* **297** 1032–39

IOP Conf. Series: Materials Science and Engineering 1167 (2021) 012015 doi:10.1088/1757-899X/1167/1/012015

Investigation on different approaches for detecting polymer content in styrene-butadiene-styrene modified asphalt

Song Xu[1], Jinming Huang[1], Yun Que[1], Yunbin Ke[2], Yong Fan[1], Canlin Zhang[1,4], and Zhengang Feng[3]

[1]College of Civil Engineering, Fuzhou University, Fuzhou 350108, China
[2]School of Civil Engineering and Architecture, Taizhou University, Taizhou 318000, China
[3]Key Laboratory of Road Structure and Material of Transport Ministry, Chang'an University, Xi'an 710064, Shaanxi, China

[4]Corresponding Author's E-mail: zhangcl@fzu.edu.cn

Abstract. This paper investigates three approaches for detecting the content of styrene-butadiene-styrene (SBS) in its modified asphalt. The physical, rheological, and structural analysis are respectively conducted by the conventional tests, dynamic shear rheology, and Fourier transform infrared spectroscopy (FTIR) tests. The results show that the fitting coefficient of the quadratic equation between the SBS content and viscosity or softening point or phase angle is very high, but these tests are time-consuming. The absorption peak at 966 cm^{-1} (the absorption peak area of 966 cm^{-1} is marked as A_{966}, the same below) is selected as the characteristic peak of SBS, and the absorption peaks of 1377 cm^{-1}, 811 cm^{-1} and 749 cm^{-1} are selected as the characteristic peak of the base asphalt in SBS modified asphalt. Comparing the applicability, accuracy and sensitivity of A_{966}/A_{1377}, A_{966}/A_{811}, and A_{966}/A_{749}, it is found that A_{966}/A_{811} is more suitable as the indicator to determine the content of SBS. FTIR can realize the rapid and accurate detection of SBS content. Furthermore, A_{966}/A_{811} has a good quadratic relationship with the performance of the modified asphalt.

1. Introduction

Asphalt has been widely utilized as aggregate binder in pavement engineering due to its good adhesion, viscoelasticity, and strength [1]. However, base asphalt is prone to be softened at high temperatures and brittle at low temperatures, which is difficult to meet the road performance requirements of modern roads. To make up the gap, polymers are used for asphalt modification. Among all the polymer modified asphalts, styrene-butadiene-styrene (SBS) modified asphalt has been acknowledged as the most excellent one, since SBS can significantly improve the high and low temperature performance, fatigue resistance, and water stability of asphalt [2]. According to the statistics, SBS modified asphalt accounts for more than 90 % of the modified asphalts in China, and asphalt modification will still be the largest terminal market for SBS block copolymers.

As is known, the SBS content is a key factor to determine the performance of the modified asphalt, so it is extremely important to obtain the accurate content of SBS in the modified asphalt for its quality assurance [3,4]. However, there is no rapid and accurate detection method to effectively determine the content of SBS at present. Because of the high cost of SBS modified asphalt, some manufacturers may exploit this loophole to gain high profits by replacing part of SBS modifier with low-cost substitute or additive, such as recycled materials [5,6].

Naskar et al. [7] evaluated the physical indicators (viscosity, softening point, penetration, and ductility) of the modified asphalt, however, this method was found to be inefficient, low accuracy, poor reproducibility, and would be easily affected by manual factors because it involved a large amount of tests, complicated detection steps, different operators [8]. Diefenderfer et al. [9] established a calibration curve of elasticity recovery and polymer content to estimate the SBS content, but the fact was that the elasticity recovery could be significantly influenced by several variables, including preparation process, additives type, and phase structure, making this approach unreliable [10]. Kosinska et al. [11] used gel permeation chromatography (GPC) to test the SBS content according to the difference in molecular weight between SBS and base asphalt, but the high molecular weight of asphaltenes, degradation and dissolution of polymers, and chemical reaction between polymers and asphalt also affected the test results. Sengoz et al. [12] studied the microscopic morphology of SBS modified asphalt by fluorescence microscope, and proposed a method to calculate the area ratio of SBS in fluorescence micrograph. But the method of threshold setting has a great uncertainty, which was greatly influenced by subjective factors and the result was highly biased. Ghavibazoo et al. [13] used the thermogravimetric analysis (TGA) test to propose a stepwise isothermal thermogravimetric fractal method by controlling the heating process. Although these methods can be used in the detection of SBS content, they still have their inherent limitations in practical application, such as high demand for equipment and environment, great influence of human factors, large deviation of detection results, as well as complex and time-consuming detection process, it is difficult to achieve a rapid, stable, accurate and easy detection.

Fourier transform infrared spectroscopy (FTIR) is the most widely implemented technology in asphalt technology. It has the advantages of simple sample pretreatment, less sample and reagent consumption, fast detection speed, high precision and mature technology [14]. It is usually adopted to analyse the aging process and anti-aging properties of asphalt [15] and compare different types of asphalt samples. FTIR can also be used to evaluate the modification mechanism of polymer modified asphalt [16-18], and to determine the presence and approximate content of polymers in modified asphalt.

Diefenderfer et al. [9] proved that FTIR spectroscopy was highly sensitive to polymer content, and the infrared peak ratio had a strong linear relationship with polymer content. Curtis et al. [19] used the FTIR to measure the content of polymer in asphalt, and obtained a linear regression model of SBS content. Sun et al. [10] found that A_{699}/A_{810} (the absorption peak area at 699 cm^{-1} is marked as A_{699}, the same below) showed a good linear relationship with SBS content. Kosinska et al. [11] used S_{965}/S_{1375} (the absorption peak at 965 cm^{-1} was marked as S_{965}, the same below) to establish a calibration curve to determine modifier content. In order to improve the accuracy of FTIR spectrum and SBS prediction model, Luo et al. [20] adopted $A_{966}/(A_{966}+A_{813})$ as the independent variable, which had the best correlation with SBS content. Those studies only used FTIR to test and verify SBS content, but there is few researches on the relationship between SBS content, infrared spectrum parameters and performance of modified asphalt.

In this paper, the advantages and disadvantages of the three detection methods are compared based on the specific test. According to the actual situation, the most appropriate modifier content detection method can be selected to realize the accurate detection of SBS content in modified asphalt, which provides guarantee for the production quality monitoring of SBS modified asphalt. The physical properties, rheology properties, and infrared spectral characteristics of SBS modified asphalt were measured, then the regression models of SBS content with the tested parameters were established by curve fitting for the purpose of quantitative determination of SBS content.

2. Experiment

2.1. Materials

The raw materials were 70# road petroleum asphalt (BA-70), SBS polymer modifier (type: 4303), sulfur (purity: 99 %), and carbon disulfide (analytical grade). The technical indicators of BA-70 and SBS are shown in Table 1 and Table 2, respectively.

Table 1. Technical indicators of BA-70.

Property	Viscosity (Pa·s)	Softening point (°C)	Penetration (0.1mm)	Ductility (cm)
Value	0.333	54.1	58	46.55

Table 2. Technical indicators of SBS.

Parameters	Unit	Value	Parameters	Unit	Value
Structure type	-	Star	300% tensile stress	MPa	2.5
Block ratio S/B	-	30/70	Tensile Strength	MPa	11.0
Oil filling rate	%	0	Elongation at break	%	750
Volatile matter	%	0.50	Tear off permanent deformation	%	12
Ash	%	0.02			

2.2. Samples preparation
The SBS modified asphalt samples were prepared in the laboratory with a high shear mixer. The weighed asphalt (350 g) was firstly heated until it became a well fluid (180 °C ~ 190 °C) in an iron container, then the SBS modifier (6 content levels, 1, 2, 3, 4, 5 and 6%, by weight of asphalt) and sulfur (two content levels, 0.1% at the SBS level of 1% ~ 4%, 0.15% at the SBS level of 5% and 6%) were added into the preheated asphalt slowly. Subsequently, 1 h shearing process was performed under a rotation rate of 4000 r/min via the high shear mixer. Finally, the blend was stirred by a mechanical stirrer for swelling 2 h. The whole preparation process was performed under the condition of oil bath heating (180 ~ 190 °C).

2.3. Physical properties test
The physical properties of asphalts, including viscosity, softening point, penetration, and ductility were tested in accordance with ASTM D 4402, D 36, D 5, and D 113, respectively.

2.4. Dynamic shear rheology (DSR) test
The DSR (MCR 102) test used strain control mode. The rheological behavior of the asphalt was determined under the strain-controlled mode by using parallel plates with 25 mm in diameter and 1 mm gap between the two parallel plates, the rheological parameters were recorded in a wide temperature scan range (from 30 °C to 120 °C with increment of 2 °C/min) at a constant frequency of 10 rad/s and a fixed strain of 0.5%. The rheological parameters, such as complex modulus (G*) and phase angle (δ) were obtained for analysis.

2.5. Fourier transform infrared spectroscopy (FTIR) test
A FTIR spectrometer with a solution of 0.125 cm^{-1} was used to obtain the infrared spectrum within the (range from 400 to 4000 cm^{-1}) of asphalt samples. The asphalt (BA-70 or SBSMA) was completely dissolved in carbon disulfide (CS_2) with 5 % concentration, then the asphalt solution was dropped onto a blank scanned KBr disk and dried for the FTIR data collection, and the infrared spectrum was analyzed by OMNIC processing software.

3. Results and discussion

3.1. Physical indicator analysis
Wu et al. [8] found that the change rules of penetration and ductility at 5 °C with SBS content were

not obvious, and the model between softening point, elastic recovery, kinematic viscosity at 135 °C and SBS content was established by grey correlation analysis method. Therefore, the viscosity and softening point are used to establish the fitting curve with SBS content in this paper. The relationship between the physical properties of SBSMA and SBS content is shown in Figure 1. Figure 1 (a) and (b) show that the addition of SBS can form a spatial network structure, which could hinder the flow of asphalt, resulting in the increase of softening point. The viscosity and softening point of SBSMA increased with the increase of SBS content, and has a high quadratic correlated with SBS content, but the accuracy of the linear relationship for accurate detection of SBS content is not enough. Moreover, the two experimental instruments are easy to obtain, but the results are easily influenced by human factors, and with respect to the efficiency, the test time is about 4 hours for six softening points and about 5 hours for six viscosity tests from moulding to the end of test.

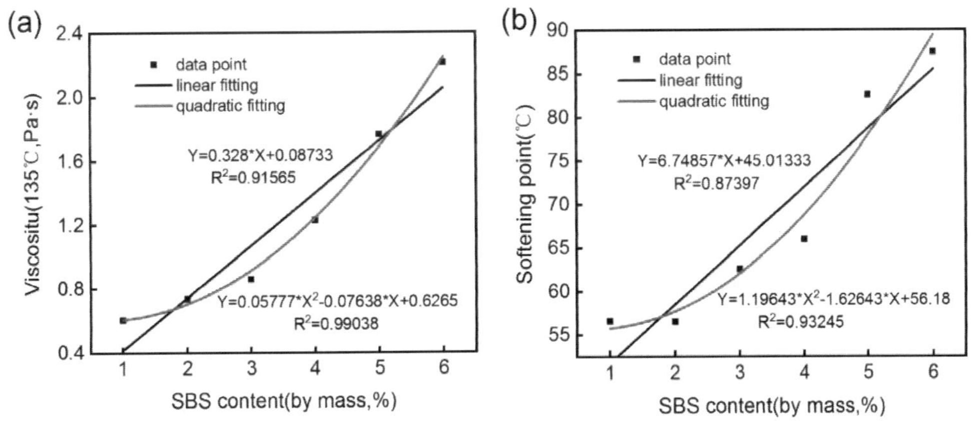

Figure 1. Relationship between physical properties of SBSMA and modifier content.

3.2. *Rheological properties analysis*

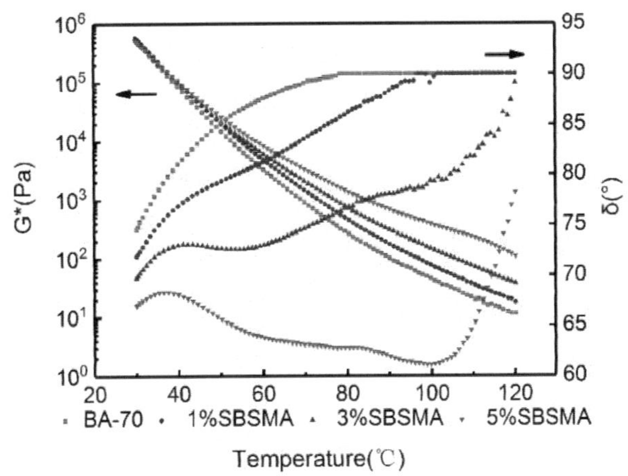

Figure 2. Plots of G* and δ versus temperature for base and modified asphalts.

Figure 2 presents the complex modulus (G*) and phase angle (δ) of BA-70 and SBSMA within the temperature range from 30°C to 120°C. It can be seen from Figure 2 that the G* curves of SBSMA are close to that of BA-70 at the lower temperature range. With the increase of temperature, the G* of both BA-70 and SBSMA decreases, with the value of SBSMA is significantly higher than that of BA-70 at the same temperature. Indicates that the incorporation of SBS polymer enhances the high temperature deformation resistance of the asphalt [21]. When the ambient temperature reaches 30 °C, the surface temperature of the asphalt pavement can reach 60 °C, so the complex modulus at 60 °C was selected as the detection index of SBS content.

From Figure 3, with the increase of SBS content, the complex modulus increases and the phase angle decreases. This is because SBS increases the elasticity of asphalt and improves the ability of asphalt to resist permanent deformation. The correlation coefficient of the phase angle (99 °C) fitting curve is the highest, which is taken as the control index of the SBS content test in the modified asphalt. Compared with the physical property test method, the rheological measurement has high degree of automation, good reproducibility and repeatability, and higher goodness-of-fit, but it has high requirements for operators, the test time is at least 7 hours for six samples' tests.

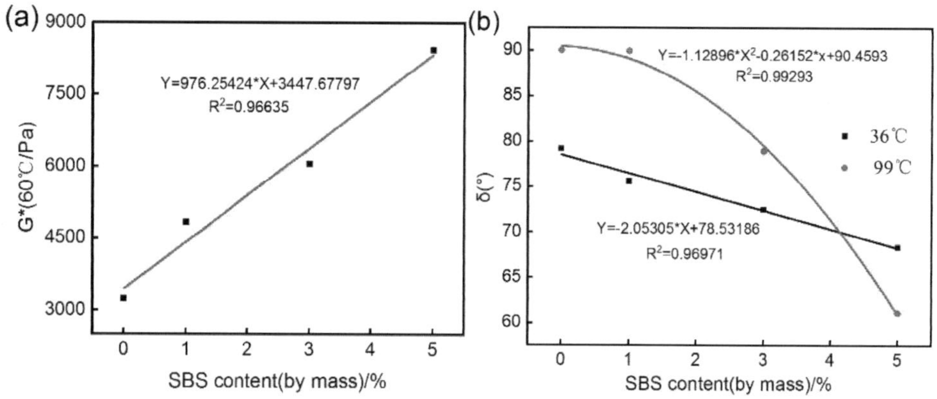

Figure 3. Relationship between rheological properties and SBS content.

3.3. FTIR analysis

The FTIR has the advantages of rapid detection, high precision, mature technology, high degree of automation and high stability [22]. Six infrared tests will take about 1.5 h, which is more time-saving than the physical and rheological properties experiments. In addition, with the development of infrared spectrometer equipment, portable infrared spectrometer and ATR-FTIR make infrared spectrum testing much easier, faster and possible to obtain infrared spectrum in the field [20]. The FTIR spectra of BA-70 and SBSMA are given in Figure 4. Figure 4 shows that no characteristic band is detected at the wavenumber greater than 3000 cm^{-1}. The BA-70 and SBSMA have obvious characteristic absorption peaks at wavenumbers of 2924 cm^{-1}, 2853 cm^{-1}, 2728 cm^{-1},1600 cm^{-1}, 1459 cm^{-1}, 1377 cm^{-1}, 1031 cm^{-1}, 811 cm^{-1},749 cm^{-1} and 723 cm^{-1}. These infrared characteristic absorption peaks represent the presence of different functional groups. The bands in the 1300 ~ 900 cm^{-1} region (called the fingerprint region) result from bone vibrations. The infrared spectrum of SBSMA has characteristic absorption peaks at 966 cm^{-1} and 699 cm^{-1}, which are significantly different from that of the BA-70 [23]. The peaks at 966 cm^{-1} and 699 cm^{-1} imply the vibration of $=CH_2$ in polybutadiene and C-H in benzene ring [24].

Since the SBS modifier does not chemically react with the BA-70, the infrared spectrum of SBSMA can be considered as a simple superposition of SBS and BA-70 [10]. The infrared spectrum of the BA-70 has almost no change at 966 cm^{-1}, with less interference than that at 699 cm^{-1} [25]. Therefore, the peak at 966 cm^{-1} has been commonly selected as a basis for qualitative and quantitative analysis of SBS [26]. In the infrared spectrum, 1377 cm^{-1} is an umbrella vibration peak of methyl ($-CH_3$), and 811 cm^{-1} is a stretching vibration of a benzene ring. Peaks at 1377 cm^{-1} and 811 cm^{-1} are obvious in both base and modified asphalts, and do not change significantly with the addition of SBS. According to the research results of Sun et al. [10] and Kosinska et al. [11], the peak of 1377 cm^{-1} and 811 cm^{-1} are selected to characterize the content of BA-70 in SBSMA.

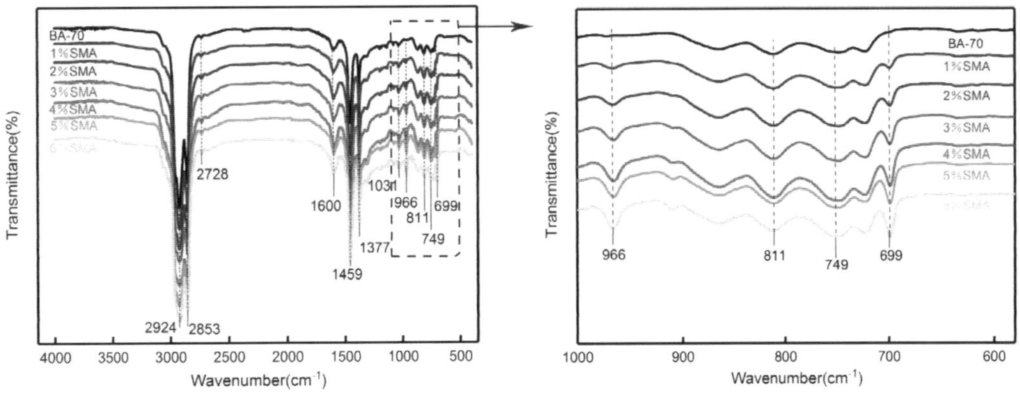

Figure 4. FTIR spectra of base asphalt and SBS modified asphalt.

During the preparation, the thickness of asphalt film to be scanned was determined by the weight of asphalt film and the diameter of the film, while the FTIR spectrum could be affected by the thickness of asphalt film. Since the concentration of asphalt solution and volume of solution dropped onto the KBr disk can be controlled, then the weight of asphalt film would be the same very time, but the diameter of the formed film is hard to be controlled. Therefore, in order to eliminate the influence of sample thickness on the measurement results, the characteristic peak area ratio of the SBS and base asphalt can be used as the parameter for the determination of SBS content [20]. According to Lambert-Beer's law, A_{966}/A_{1377} and A_{966}/A_{811} are used as the quantitative analysis index of SBS content. It can also be found that A_{966}/A_{749} (the peak at 749 cm^{-1} is attributed to the bending vibration of the aromatic branched chain) also changed linearly with SBS content. So the fitting curve between A_{966}/A_{1377}, A_{966}/A_{811} or A_{966}/A_{749} and SBS content was established to determine the content of SBS in modified asphalt, and the fitting results are shown in Figure 5. It can be seen that the correlation coefficient is greater than 0.98.

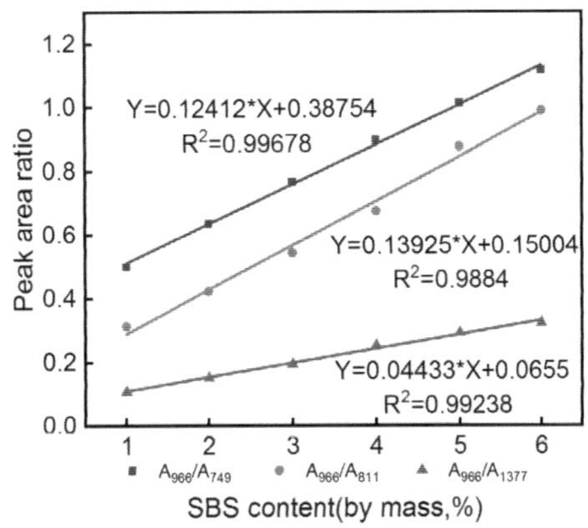

Figure 5. The fitting curve of peak area ratio and SBS content.

The base asphalt and SBS used in SBS modified asphalt market are inconsistent with those in this paper, so it is necessary to verify the universality of FTIR detection method. Two different SBS modified asphalts were prepared by using another base asphalt (A-70) and SBS (SBS 1303): SBS 1303 modified BA-70 (V1) and SBS 4303 modified A-70 (V2). The newly prepared modified asphalt was analyzed by infrared spectroscopy, and A_{966}/A_{1377}, A_{966}/A_{811}, and A_{966}/A_{749} were substituted into the prediction equations presented in Figure 5, respectively. The prediction results are shown in Table 3.

The maximum absolute error of the predicted value after changing the type of SBS or base asphalt is 2.17, far exceeding the acceptable range of error. Therefore, the prediction equation with 749 cm^{-1} as the reference characteristic peak has a small application range. When A_{966}/A_{811} is used as the dependent variable, the linear correlation between A_{966}/A_{811} and the content of the modifier is relatively low. However, it is most sensitive to changes in modifier content. After changing the type of SBS or base asphalt, the absolute error of the prediction results is controlled within 0.62. It indicates that A_{966}/A_{811} is suitable for different types of modified asphalt. With A_{966}/A_{1377} as the dependent variable, the linear correlation between the peak area ratio and SBS is good, and the absolute error of the prediction results of the modifier content of newly prepared modified asphalt is controlled within 1.08. As shown in Figure 4, there is no interference absorption peak near the absorption peak at 1377 cm^{-1}, 811 cm^{-1} and 749 cm^{-1}, with high identification. The peak shape at 1377 cm^{-1} is sharp and the intensity is high, while the intensity of 811 cm^{-1} and 749 cm^{-1} absorption peaks is weak, about 1/4 of 1377 cm^{-1}. In summary, comparing the applicability and accuracy of A_{966}/A_{1377}, A_{966}/A_{811} and A_{966}/A_{749}, it is more suitable to choose A_{966}/A_{811} as the indicator to determine the content of SBS. The prediction equation is Y = 0.13925*X + 0.15004 (Y = A_{966}/A_{811}, X = SBS content). FTIR can achieve fast and accurate detection of SBS content in modified asphalt.

Table 3. Prediction results of SBS content in different fitting curves.

Dependent variable	Predicted value (%)					
	A_{966}/A_{1377} Y = 0.12412*X + 0.38754		A_{966}/A_{811} Y = 0.13925*X + 0.15004		A_{966}/A_{749} Y = 0.04433*X + 0.0655	
SBS content /%	V1	V2	V1	V2	V1	V2
1	1.33328	0.26607	1.08975	0.96631	1.35488	0.90442
2	1.92695	1.70037	1.94852	1.94190	2.40760	3.31363
3	2.42557	3.75884	3.09393	3.22570	2.49849	5.09753
4	4.49502	4.24146	4.25554	4.14260	5.46804	6.17409
5	4.92126	5.48705	4.94039	5.43610	5.98056	6.97751
6	6.13274	7.07836	6.22933	6.39145	7.32336	7.86206

Table 4 shows the theoretical values of 70# road petroleum modified asphalt modifiers with different fitting curves. Diefenderfer [9], Kosinska [11], Sun [10], and Luo [20] have established the relationship between infrared spectra and SBS content, the fitting equations are Y_1, Y_2, Y_3, Y_4, respectively. Diefenderfer and Kosinska used S_{965}/S_{1375} to calculate the SBS content, while Sun and Luo used A_{699}/A_{810} and A_{966}/A_{1377} to calculate the SBS content respectively.

In this paper, the fitting curve established with A_{966}/A_{811} as the dependent variable is Y, and A_{966}/A_{811} of SBSMA is substituted into the equation. As shown in Table 4, there is a slight difference between the calculated and actual values of the six samples, and the absolute error is controlled within 0.23, which is an acceptable range. In addition, when S_{965}/S_{1375} and A_{699}/A_{810} are substituted into the equation, the calculated value of SBS content is significantly different from the actual value, and the absolute error is greater than 0.5.

Moreover, modified asphalt studied by Diefenderfer is obtained by adding SBS into the base asphalt and shearing it for 2 h at 185 °C and 1200 r/min. The modified asphalt used by Kosinska is field samples collected in the Gdansk area at different time during the road construction. These illustrate that the sample preparation method and asphalt aging degree have a significant influence on the SBS content detection. Sun calculated the SBS content by using the laboratory-made modified asphalt peak area ratio A_{699}/A_{810}. However, the 810 cm-1 represents the stretching vibration of the benzene ring in asphalt, and the infrared spectrum of base asphalt is unstable at 699 cm^{-1}, leading to the inaccuracy results. The absolute error of Y_4 and Y is less than 0.59, so the influence of SBS and stabilizer types on SBS content can be neglected. Hence, it is more accurate to estimate SBS content using the peak area at 966 cm^{-1}.

The above analysis shows that the types of SBS and stabilizers have little effect on the determination of SBS content. As long as the preparation method and aging degree of the standard sample and the sample to be tested are the same, the SBS content can be accurately detected. Infrared spectroscopy can realize the supervision of the amount of modifier in SBSMA, and provide reliable guarantee to ensure the construction quality of asphalt pavement, extending the service life of pavement and reducing the cost of pavement maintenance.

Table 4. Comparison of several different regression equations

Source	This article		Literature [9]		Literature [11]		Literature [10]		Literature [20]	
Equation	$Y = 0.13925*X + 0.15004$		$Y_1 = 0.072*X_1 + 0.016$		$Y_2 = 0.031*X_2 + 0.0283$		$Y_3 = 0.1048*X_3 - 0.0763$		$Y_4 = 0.046*X_4 + 0.0423$	
Y value	A_{966}/A_{811}		S_{966}/S_{1377}		S_{966}/S_{1377}		A_{699}/A_{810}		A_{966}/A_{1377}	
SBS (%)	X	Y	X	Y	X	Y	X	Y	X	Y
1	1.177	0.314	0.875	0.079	1.645	0.079	2.150	0.149	1.450	0.109
2	1.953	0.422	1.319	0.111	2.677	0.111	2.837	0.221	2.385	0.152
3	2.829	0.544	1.889	0.152	4.000	0.152	3.886	0.331	3.320	0.195
4	3.777	0.676	2.403	0.189	5.194	0.189	4.841	0.431	4.580	0.253
5	5.228	0.878	3.222	0.248	7.097	0.248	6.596	0.615	5.428	0.292
6	6.032	0.990	3.764	0.287	8.355	0.287	7.541	0.714	6.102	0.323
	Absolute error (%)									
1	0.177		-0.125		0.645		1.150		0.450	
2	-0.047		-0.681		0.677		0.837		0.385	
3	-0.171		-1.111		1.000		0.886		0.320	
4	-0.223		-1.597		1.194		0.841		0.580	
5	0.228		-1.778		2.097		1.596		0.428	
6	0.032		-2.236		2.355		1.541		0.102	

4. Conclusions

The content of modifier is crucial to the performance of modified asphalt. In this paper, the physical properties, rheological properties and infrared tests of SBSMA were carried out to quantitatively analyze the content of SBS in the modified asphalt. The main conclusions are as follows:

(1) FTIR can detect the polymer content of SBS modified asphalt quickly and accurately. The regression model of A_{966}/A_{811} (Y) and SBS content (X) is established as follows: $Y = 0.13925*X + 0.15004$. The correlation coefficient can reach 0.9884. Two kinds of modified asphalt are used to verify the formula, and the absolute errors between the predicted results and the actual results are within the acceptable range. Compared with the fitting curves of literatures, it is found that the aging degree and sample preparation method have great influence on the determination of SBS content, while the influence of SBS type and stabilizer type can be ignored.

(2) The linear correlation between SBS content and viscosity or softening point or phase angle (99 °C) is not accurate enough to accurately detect SBS content. The quadratic function fitting correlation is high, but this method is time-consuming. Moreover, due to the low degree of automation, the accuracy of viscosity and softening point results are easily affected by human factors, and the test reproducibility is poor. The DSR test has a high degree of automation, so the accuracy of the method based on rheological parameters is higher than that based on physical properties. However, DSR test

has the high requirements on personnel and environment, and the test time is relatively long, so it is difficult to apply in practical engineering.

Acknowledgments
This work was supported by the National Natural Science Foundation of China (51708121) and the Open Fund of Key Laboratory of Road Structure and Material of Transport Ministry, Chang'an University (300102219522). The authors gratefully acknowledge their financial support.

References
[1] Singh B, Kumar P. Effect of polymer modification on the ageing properties of asphalt binders: Chemical and morphological investigation, *Construction and Building Materials*, 2019, **205**: 633-641.
[2] Xiao F, Amirkhanian S, Wang H, Hao P. Rheological property investigations for polymer and polyphosphoric acid modified asphalt binders at high temperatures, *Construction and Building Materials*, 2014, **64**: 316-323.
[3] Al-Hadidy A I, Tan Y. Effect of styrene-butadiene-styrene on the properties of asphalt and stone-matrix-asphalt mixture, *Journal of Materials in Civil Engineering*, 2011, **23**(4): 504-510.
[4] Hou X, Lv S, Chen Z, Xiao F. Applications of fourier transform infrared spectroscopy technologies on asphalt materials, *Measurement*, 2018, **121**: 304-316.
[5] Cai L, Shi X, Xue J. Laboratory evaluation of composed modified asphalt binder and mixture containing nano-silica/rock asphalt/SBS, *Construction and Building Materials*, 2018, **172**: 204-211.
[6] Ahmedzade P, Fainleib A, Günay T, Grygoryeva O. Modification of bitumen by electron beam irradiated recycled low density polyethylene, *Construction and Building Materials*, 2014, **69**: 1-9.
[7] Naskar A K, De S K, Bhowmick A K, Pramanik P K, Mukhopadhyay R. Mukhopadhyay, Characterization of ground rubber tire and its effect on natural rubber compound, *Rubber Chemistry and Technology*, 2000, **73**(5): 902-911.
[8] Wu J M, Xue-Li Z, Zhong-Da C, Jie L, Hong-Liang Z. Test method of SBS content in modified asphalt, *Journal of Traffic & Transportation Engineering*, 2012, **26**(2): 24-26.
[9] Diefenderfer S D. Detection of polymer modifiers in asphalt binder, *Virginia Transportation Research Council*, 2006.
[10] Sun, D, Zhang L. Quantitative determination of SBS content in SBS modified asphalt. *Journal of Building Materials*, 2013, **16**(1): 180-184. (in Chinese)
[11] Kosinska J, Boczkaj G, Galezowska G, Podwysocka J, Przyjazny A, Kaminski M. Determination of modifier contents in polymer-modified bitumens and in samples collected from the roads using high-performance gel permeation/size-exclusion chromatography, *Road Materials & Pavement Design*, 2016, **17**(3): 547-562.
[12] Sengoz B, Isikyakar, G. Analysis of styrene-butadiene-styrene polymer modified bitumen using fluorescent microscopy and conventional test methods, *Journal of hazardous materials*, 2008, **150**(2): 424-32.
[13] Ghavibazoo A, Abdelrahman M, Ragab M. Mechanism of crumb rubber modifier dissolution into asphalt matrix and its effect on final physical properties of crumb rubber–modified binder, *Transportation Research Record*, 2013, **2370**(1): 92-101.
[14] Hirri A, Bassbasi M, Souhassou S, Kzaiber F, Oussama A. Prediction of polyphenol fraction in virgin olive oil using mid-infrared attenuated total reflectance attenuated total reflectance accessory–mid-infrared coupled with partial least squares regression, *International Journal of Food Properties*, 2016, **19**(7): 1504-1512.
[15] Nivitha M R, Prasad E, Krishnan J M. Ageing in modified bitumen using FTIR spectroscopy, *International Journal of Pavement Engineering*, 2016, **17**(7): 565-577.
[16] Zhang F, Hu C. The research for structural characteristics and modification mechanism of crumb

rubber compound modified asphalts, *Construction and Building Materials*, 2015, **76**: 330-342.
[17] Rasool R T, Wang S, Zhang Y, Li Y, Zhang G. Improving the aging resistance of SBS modified asphalt with the addition of highly reclaimed rubber, *Construction and Building Materials*, 2017, **145**: 126-134.
[18] Jiang Z, Easa S, Hu C, Zheng X. Understanding damping performance and mechanism of crumb rubber and styrene-butadiene-styrene compound modified asphalts, *Construction and Building Materials*, 2019, **206**: 151-159.
[19] Curtis C W, Hanson D I, Shou T C, Shieh G J, Ling M. Quantitative determination of polymers in asphalt cements and hot-mix asphalt mixes, *Transportation Research Record*, 1995, **1488**: 52-61.
[20] Luo S, Tian J, Liu Z, Lu Q, Zhong K, Yang X. Rapid determination of styrene-butadiene-styrene (SBS) content in modified asphalt based on Fourier transform infrared (FTIR) spectrometer and linear regression analysis, *Measurement*, 2020, **151**: 107204.
[21] Wu S, Pang L, Mo L, Chen Y, Zhu G. Influence of aging on the evolution of structure, morphology and rheology of base and SBS modified bitumen, *Construction and Building Materials*, 2009, **23**(2) 1005-1010.
[22] Haas J, Mizaikoff B. Advances in mid-infrared spectroscopy for chemical analysis, *Annu Rev Anal Chem (Palo Alto Calif)*, 2016, **9**(1): 45-68.
[23] Sheng X, Wang M, Xu T, Chen J. Preparation, properties and modification mechanism of polyurethane modified emulsified asphalt, *Construction and Building Materials*, 2018, **189**: 375-383.
[24] Xu S, Yu J, Hu C, Qin D, Xue L. Laboratory evaluation of rejuvenation effect of reactive rejuvenator on aged SBS modified bitumen, *Materials and Structures*, 2017, **50**(6): 233..
[25] Yan C, Xiao F, Huang W, Lv Q. Critical matters in using attenuated total reflectance fourier transform infrared to characterize the polymer degradation in Styrene–Butadiene–Styrene-modified asphalt binders, *Polymer Testing*, 2018, **70**: 289-296.
[26] Ye Z C, Chen D, Ling C, Guan F Y. Quantifying SBS content in modified asphalt using Fourier transform infrared spectroscopy, *Advanced Materials Research*, 2015, **1065-1069**: 691-695.

IOP Conf. Series: Materials Science and Engineering 1167 (2021) 012016 doi:10.1088/1757-899X/1167/1/012016

Evaluation of 3-methacryloxypropyltrimethoxysilane organic layered double hydroxide on the aging resistance of bitumen

Canlin Zhang[1,2,4], Minxuan Chen[1,2], Ting Wang[3], Jun He[1,2], Guochuan Chen[4], Song Xu[1,2,5] and Changbin Hu[1,2]

[1]College of Civil Engineering, Fuzhou University, Fuzhou 350108, P.R. China
[2]Fujian Provincial University Engineering Research Center for Advanced Civil Engineering Materials, Fuzhou University, Fuzhou 350108, P.R. China
[3]Testing Center of Fuzhou University, Fuzhou 350108, P.R. China
[4]School of Advanced Manufacturing, Fuzhou University, Fuzhou 350108, P.R. China

[5]Corresponding Author's E-mail: xusong@fzu.edu.cn

Abstract. To enhance and lengthen the anti-aging performance and service lifetime of bitumen, 3-methacryloxypropyltrimethoxysilane grafted layered double hydroxides (LDHs) was synthesized and exploited to modify bitumen. The contact angle of organic LDHs (OLDHs) increased from 17.8° to 39.2°, the hydrophilic of OLDHs obviously decreased in comparison with LDHs, which contributed to disperse more uniformly in bitumen. Furthermore, the stability of LDHs in bitumen became better after 3-methacryloxypropyltrimethoxysilane modification. After aging, the penetration and ductility of bitumen markedly declined, correspondingly, the softening point and viscosity remarkably rose, the physical performance of bitumen were deteriorated seriously. The addition of LDHs and OLDHs could restrain the properties degradation after aging, enhance the anti-aging of bitumen, and OLDHs exhibited much more excellent, because of the outstanding stability in bitumen.

1. Introduction

Bitumen has become the widely application material in pavement construction, because of its outstanding performance, such as driving comfort, low noise and convenient maintenance, etc [1-3]. Unfortunately, exposing the environment (heat, oxygen, ultraviolet light), bitumen will be aged, which would deteriorate the performance of bitumen, ultimately shorten the service lifetime of bitumen pavement [4,5]. Hence, it is of great practical significance to obtain bitumen with preeminent anti-aging performance.

To enhance the anti-aging performance of bitumen, some antiaging agents have been added. Feng et al [6,7] found that the anti-aging performance of bitumen could be perfected by ultra-violet (UV) absorber, but the effectiveness of UV absorber was selective, the specific UV absorber would only ameliorate the anti-aging of certain kinds of bitumen, the mis-selection would accelerate the aging of bitumen. Zhang et al [4,8,9] noticed that nanoparticles could exhibit better performance in ameliorating the anti-aging of bitumen, but the dispersity of nanoparticles in bitumen was bad, which severely limited the application of nanoparticles in bitumen. In addition, layered silicate could also ameliorate the aging resistance of bitumen, but layered silicate could not exist stably in bitumen [10].

Layered double hydroxides (LDHs) is a kind of inorganic nanomaterial with multi-layers structure [11,12]. Due to the unique structure, it could reflect and absorb ultraviolet light, and meanwhile obstruct oxygen and heat into bitumen. Therefore, LDHs would be a better choice to prepare bitumen with excellent aging resistance. Recently, LDHs has been applied in the bitumen modification, and the results indicated that LDHs could enhance the anti-aging to a certain extent [13,14]. However, it is noteworthy that the surface of LDHs distributes a mass of hydroxide radicals, resulting in the heterogeneous dispersion in bitumen, eventually the enhancement of LDHs on the anti-aging could not fully exert. Hence, it is desiderating to make LDHs disperse better in bitumen.

In this paper, silane coupling agent grafted LDHs (OLDHs) was synthesized and used to kaizen the anti-aging of bitumen, the hydrophilic of OLDHs and LDHs, and the stability in bitumen was analyzed. The effect of LDHs before and after grafting modification on the aging performance of bitumen was investigated.

2. Materials and methods

2.1. Materials

The physical properties of pristine bitumen were shown in Table 1, which was provided by SK Corp., Ltd. LDHs were supplied from the company named Beijing Tech-layer, Ltd. Silane coupling agent (3-methacryloxypropyltrimethoxysilane) was produced by Aladdin Industrial Corp., Ltd.

Table 1. Physical property of pristine bitumen.

Items		Parameters
Penetration (25 °C, 0.1mm)		73
Softening point (°C)		48.8
Ductility (cm)	15 °C	>150
	10 °C	16.5
Viscosity (135°C, Pa·s)		0.49

2.2. Preparation of OLDHs

Primarily, LDHs and 3-methacryloxypropyltrimethoxysilane were dispersed in ethanol-solution. Then, under the nitrogen protection and rapid stirring, the solutions reacted at 70 °C for 0.5 h. Afterwards, the LDHs solution was washed repeatedly and dried at 80 °C for 12 h.

2.3. Preparation of OLDHs modified asphalt

Firstly, pristine bitumen was melted to flow at 150±5 °C. Then, OLDHs and LDHs was added into binder. Subsequently, the compound was under high-speed stirring at 170±5 °C for 1.5 h. For the sake of conciseness, blank bitumen, bitumen modified by OLDHs and LDHs were shorted for MA, OLMA and LMA, respectively.

2.4. Static contact angle test

The hydrophilia of LDHs before and after modification was measured by Contact Angle Meter (CAM, SL 200C, USA).

2.5. Aging method

The molten bitumen was fed into metal plates. Then, the specimens were put in thin film oven to simulate short-term thermal oxygen aging. After that, the specimens were transferred to UV aging oven, whose light intensity is 2500 μW/cm^2, at 60 °C for 7 days.

2.6. Physical properties tests

The physical properties of all binder specimens were measured according to ASTM D5976, D5, D36, D113 and D4402, respectively.

3. Results and discussion

3.1. Hydrophilicity analysis of OLDHs

The contact angles of LDHs and OLDHs are portrayed in Fig. 1. The numerical magnitude of contact angle represents the hydrophilicity of LDHs. The larger the contact angle, the less the hydrophilicity. Fig. 1(a) showed that the contact angle of LDHs was 17.8°, it indicated that LDHs had the better hydrophilicity, due to the plentiful polar groups (hydroxyl radicals) in LDHs, which was just the reason of the easy agglomeration of LDHs particle and difficult dispersion in bitumen. In contrast, the contact angle of OLDHs increased to 39.2°, which was attributed to the reduced free hydroxyls and the newly grafted organic group in LDHs after 3-methacryloxypropyltrimethoxysilane modification. The hydrophilicity of LDHs was weakened after 3-methacryloxypropyltrimethoxysilane surface organic modification, which indicated that 3-methacryloxypropyltrimethoxysilane organic modification was helpful for enhancing the dispersibility of LDHs in hydrophobic bitumen.

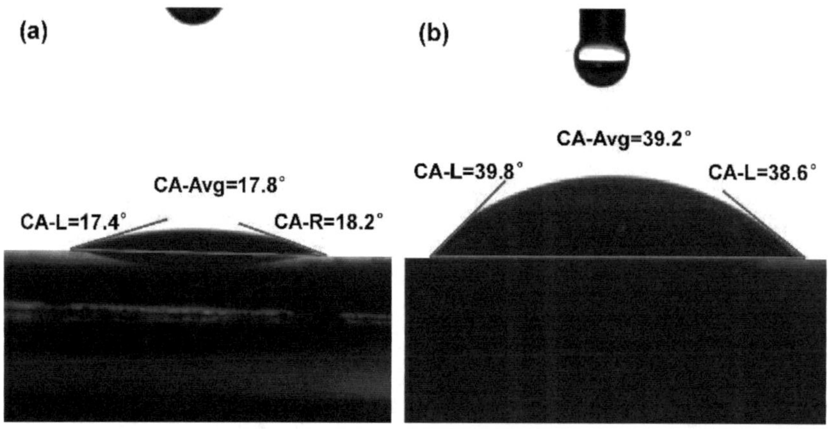

Figure 1. The contact angles of (a) LDHs and (b) OLDHs

3.2. Storage stability

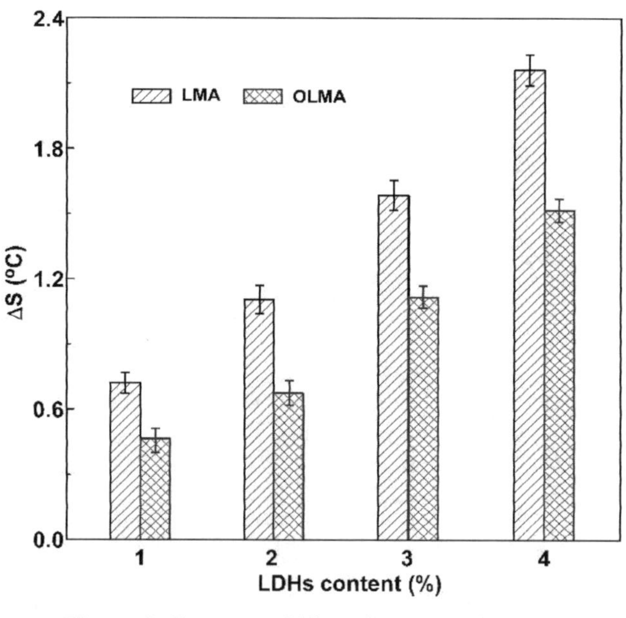

Figure 2. Storage stability of LMA and OLMA

Bitumen with better storage stability is significance for the anti-aging of bitumen. Fig. 2 showed the storage of LMA and OLMA. It could be found that the softening point difference (△S) of all samples was increased with the LDHs increasing, which verified that LDHs could not evenly and stably scatter in bitumen. Compared with LMA, the △S of OLMA decreased significantly, and the tendency became more obvious with the LDHs content from 1% to 4%. The result manifested that 3-methacryloxypropyltrimethoxysilane organic modification was helpful for improving the stability of LDHs in bitumen.

3.3. *Physical properties*

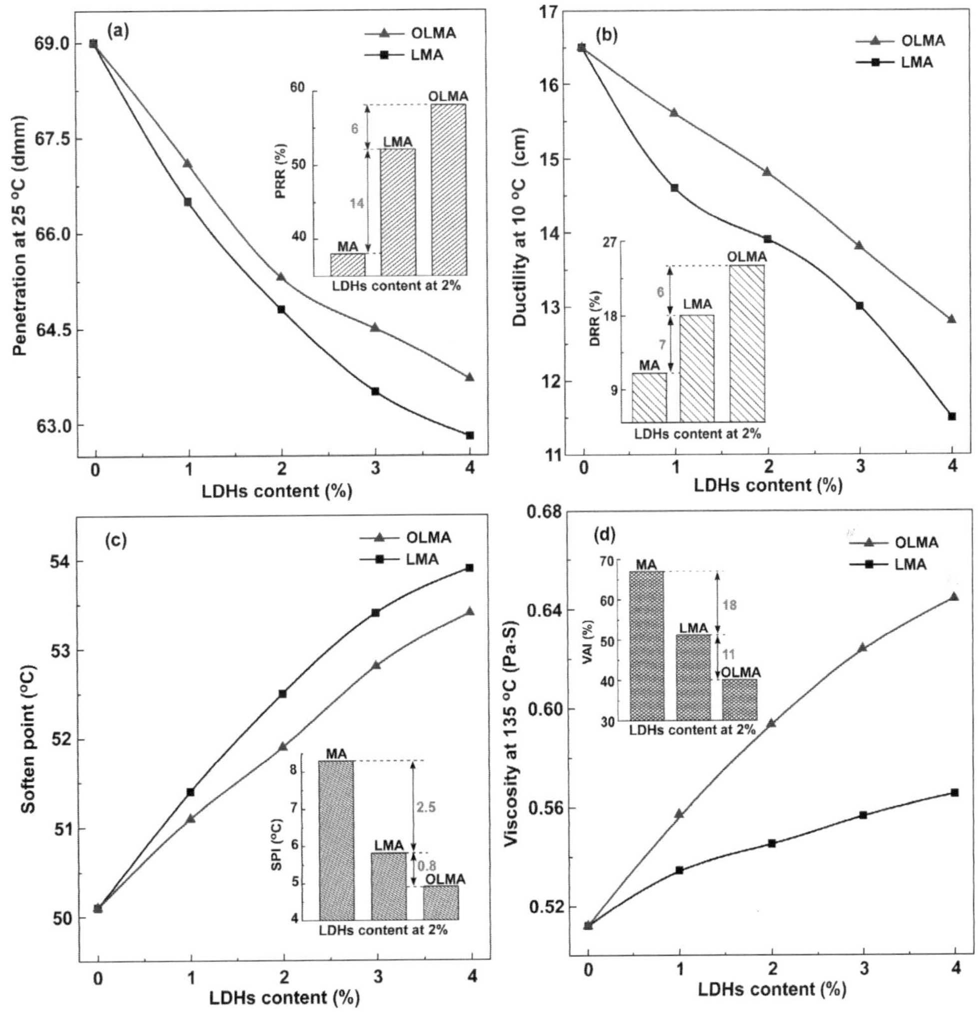

Figure 3. Physical properties of LMA and OLMA before and after ultraviolet aging: (a) penetration at 25 °C, (b) softening point, (c) ductility at 10 °C, (d) viscosity at 135 °C

The physical performance of all samples are plotted in Fig. 3. It could be found that LDHs reduced the penetration and ductility, enlarged the softening point and viscosity. Furthermore, the effect became more distinct with LDHs increase. As inorganic powder, adding LDHs will cause the elastic behaviour of bitumen increase. Notably, compared with LDHs, OLDHs had less impact on the penetration, ductility and softening point, while had a greater effect on viscosity, which was attributed to the preferable stability in bitumen.

In order to appraise the improvement of OLDHs on the anti-aging, the aging indexes were used in this paper. The computational equations of these aging indexes were as follows:

$$\text{PRR}=\text{Penetration after aging/Penetration before aging} \quad (1)$$

$$\text{SPI}=\text{Softening point after aging-Softening point before aging} \quad (2)$$

$$\text{DRR}=\text{Ductility before aging/Ductility after aging} \times 100 \quad (3)$$

$$\text{VAI}=(\text{Viscosity after aging-Viscosity before aging})/\text{Viscosity before aging} \times 100 \quad (4)$$

The aging indexes of all specimens are depicted in Fig. 3 (embedded figure). The physical properties of bitumen were deteriorated seriously after aging, penetration and ductility markedly reduced, correspondingly, softening point and viscosity remarkably increased. Compared with MA (PRR of 43% and DRR of 12%), the PRR and DRR of LMA were 57% and 19%, OLMA exhibited more prominent (PRR of 57% and DRR of 19%), it showed that OLDHs or LDHs could retard the penetration and ductility decrease of bitumen during aging, especially OLDHs. Correspondingly, SPI and VAI of MA, LMA and OLMA exhibited different rule, the size order of SPI and VAI was MA > LMA > OLMA, it indicated that OLDHs or LDHs could slow down the effect of aging on softening point and viscosity. The result indicated that LDHs could strengthen the anti-aging ability of bitumen, alleviate the property deterioration during aging, furthermore, the effect of LDHs has been further heightened after 3-methacryloxypropyltrimethoxysilane modification.

4. Conclusions

3-methacryloxypropyltrimethoxysilane grafted LDHs was synthesized and used to enhance the anti-aging property of bitumen. Based on the experimental results, some conclusions could be drawn as follows:

(1) In comparison with LDHs, the hydrophilia of modified LDHs significantly decreased, which made LDHs dispersing more uniformly in bitumen.

(2) The stability of OLDHs in bitumen was better than that of LDHs, and the tendency became more obvious with the LDHs increasing, which was due to the newly introduced organic groups into LDHs after 3-methacryloxypropyltrimethoxysilane organic modification.

(3) The physical properties of bitumen became worse after aging, which was alleviated by adding LDHs, and the anti-aging ability of bitumen has been improved. The enhancement of LDHs has been strengthened after 3-methacryloxypropyltrimethoxysilane organic modification.

Acknowledgments

This work is supported by National Natural Science Foundation of China (52008113), Educational Research Projects for Young and Middle-Aged Teachers of Education Department of Fujian Province (JAT190047), Open Fund of Key Laboratory of Road Structure and Material of Ministry of Transport, Chang'an University (300102210509), State Key Laboratory of Special Functional Waterproof Materials (SKLW2019012) and Undergraduate scientific research training program of Fuzhou University (051801425). The authors gratefully acknowledge their financial support.

References

[1] Lu X, Isacsson U. 1998. Chemical and rheological evaluation of ageing properties of SBS polymer modified bitumens, *Fuel.* **77**(9-10) 961-72.

[2] Tauste R, Moreno-Navarro F, Sol-Sánchez M, Rubio-Gámez MC. 2018. Understanding the bitumen ageing phenomenon: a review, *Construction and Building Materials.* **192** 593-609.

[3] Xu X, Yu JY, Xue LH, He BY, Du W, Zhang H, Li Y. 2018. Effect of reactive rejuvenating system on physical properties and rheological characteristics of aged SBS modified bitumen, *Construction and Building Materials.* **176** 35-42.

[4] Zhang HL, Chen ZH, Xu GQ, Shi CJ. 2018. Evaluation of aging behaviors of asphalt binders

through different rheological indices, *Fuel.* **221** 78-88.
[5] Zhu J, Birgisson B, Kringos N. 2014. Polymer modification of bitumen: advances and challenges, *European Polymer Journal.* **54** 18-38
[6] Feng ZG, Yu JY, Xue LH, Sun YB. 2013. Rheological and aging properties of ultraviolet absorber/styrene-butadiene-styrene-modified bitumens, *Journal of Applied Polymer Science.* **128** 2571-77.
[7] Feng ZG, Cai FJ, Yao DD, Li XJ. 2021. Aging properties of ultraviolet absorber/SBS modified bitumen based on FTIR analysis, *Construction and Building Materials.* **273** 121713.
[8] Zhang HL, Zhu CZ, Chen ZH. 2017. Influence of multi-dimensional nanomaterials on the aging behavior of bitumen and SBS modified bitumen, *Petroleum Science and Technology.* **35** 1931-37.
[9] Zhang HL, Zhang DM. 2015. Effect of different inorganic nanoparticles on physical and ultraviolet aging properties of bitumen, *Journal of Materials in Civil Engineering.* **27** 04015049.
[10] Zhang HL, Zhu CZ, Yan KZ, Yu JY. 2017. Effect of rectorite and its organic modification on properties of bitumen, *Journal of Materials in Civil Engineering.* **27** C4014002.
[11] Li W, Yan DP, Gao R, Lu J, Wei M, Duan X. 2013. Recent advances in stimuli-responsive photofunctional materials based on accommodation of chromophore into layered double hydroxide nanogallery, *Journal of Nanomaterials.* **2013** 586462.
[12] Li CM, Wei M, Evans DG, Duan X. 2015. Recent advances for layered double hydroxides (LDHs) materials as catalysts applied in green aqueous media, *Catalysis Today.* **247** 163-169.
[13] Xu S, Li L, Yu JY, Zhang CL, Zhou J, Sun YB. 2015. Investigation of the ultraviolet aging resistance of organic layered double hydroxides modified bitumen, *Construction and Building Materials.* **96** 127-134.
[14] Zhang CL, Yu JY, Xu S, Xue LH, Cao ZL. 2016. Influence of UV aging on the rheological properties of bitumen modified with surface organic layered double hydroxides, *Construction and Building Materials.* **123** 574-580.

Viscoelastic behaviors of high-viscosity rubberized bitumen containing waste cooking oil

Xiong Xu[1], Qiaoyun Wu[1], Weili Shen[1], Shukai Cheng[1], and Xuyong Chen[1,2]

[1]School of Civil Engineering and Architecture, Wuhan Institute of Technology, Wuhan, China

[2]Corresponding Author's E-mail: Chenxy@wit.edu.cn

Abstract. With the modification of crumb rubber (CR), the viscosity and elasticity of asphalt binder will be significantly enhanced. This study used waste cooking oil (WCO) to adjust the chemical fractions in the prepared CR modified asphalt binder (CRMA) for the purpose of reaching a better viscoelastic behavior. Further, the viscoelastic behavior of CRMA incorporated with 4% WCO was investigated and evaluated by constructing reliable master curves from outputs of the classical Christensen-Anderson model. The findings obtained indicated that the use of 4%WCO into CRMA binder just causes a limited decrease of the complex modulus, and in this case, the resulting binder still remains applicable to stand with the resistance to deformation from outside repeated forces; both raising temperature and incorporating WCO can be able to produce similar effects on the viscoelastic transition of CRMA binder approaching to higher frequency; once the environmental condition surpasses the viscoelastic transition temperature of CRMA binder, or WCO is added, the viscous behavior will be switched to a lower temperature; the Christensen-Anderson model is verified available and supportively recommended for viscoelastic analyses of binders.

1. Introduction

As of now, the waste materials, such as vehicle tire rubber, waste oil, plastic, and other kind of discarded landfilled categories, have been the severely pollutant issues concerned by international society [1,2]. For those which may be recycled and utilized, it is now being of great urgency to follow up for the purpose of environmental protection and resource cycling.

Crumb rubber (CR) from tires, as a value-added material, has been widely applied in modification of bituminous materials because of its outstanding high-temperature properties, elasticity, and aging resistance [3,4]. Some publications indicate that the overall properties of rubberized bitumen can be improved as CR reaches an appropriate incorporation [5,6]. Although, however, they work well in those aspects, the resulting rubberized binder is not able to be easily blended due to its deteriorated workability. Waste cooking oil (WCO) is a common post-consumer liquid waste that majorly comes from restaurants. Considering its low-viscosity and organic components, it has been recycled and used to improve the workability of high-viscosity bitumen and restore the properties of aged binders for rejuvenation [7, 8].

Viscoelastic behavior is an important characteristic to identify the rheological performances of asphalt binder for determining its availability in pavement under comprehensive conditions of temperature, frequency, and stress [9]. The technical indexes of storage modulus (G′) and loss modulus (G″), also known as the elastic and viscous modulus, refer to the elastic recoverable component and viscous irreversible component of material, respectively. Thus, great importance needs be closely attached to the intercrossing curve point of G′ and G″ over temperatures or frequencies to understand the viscoelasticity of binders.

Taking the mentioned-above basis, this study utilizes appropriate percentage of WCO to improve the workability of high-viscosity rubberized binder, and subsequently investigates the viscoelastic behavior of CRMA incorporating WCO by constructing reliable master curves from the outputs of classical Christensen-Anderson model.

2. Christensen-Anderson model

In 1992, Christensen and Anderson proposed a typical mathematical model (equation 1) to develop master curves for fully evaluating viscoelastic properties of asphalt binder [10], which was confirmed and frequently utilized to investigate rheological characteristics of binders in asphalt field.

$$|G^*| = G_g\left[1 + \left(\frac{w_c}{w_r}\right)^{(\log 2)/R}\right]^{-R/(\log 2)} \tag{1}$$

where, G* refers to the complex modulus, Pa; G_g refers to the glass modulus, equal to 1GPa, Pa; w_c and wr refer to the crossover frequency and reduced frequency at reference temperature, respectively, rad/s; R refers to the rheological index. Among these, G_g, w_c, and R are the fitting parameters to be determined in processing.

In respect of this model, it is of great importance to focus on wr that needs be assured by equation 2 as below. Subsequently, log a(T), the shift factor, to be determined by William-Landel-Ferry (WLF) model [11], is showed in equation 3.

$$w_r = w \times 10^{\log a(T)} \tag{2}$$

$$\log a(T) = \frac{-C_1(T-T_r)}{C_2+T-T_r} \tag{3}$$

where, w refers to the test frequency, rad/s; T refers to the test temperature, °C; T_r refers to the reduced temperature, °C; C_1 and C_2 are the fitting parameters to be determined according to optimized results.

3. Materials and methodology

3.1. Materials preparation

High-viscosity rubberized bitumen (CRMA) and waste cooking oil (WCO) were supplied by a local asphalt factory and restaurant, respectively. For the purpose of improving the workability of virgin CRMA, 4%WCO, by weight of virgin binder, was incorporated and continuously mixed at 160~170°C for 1h. To better understand the significance of this work, some of the performance indexes are shown in Table 1, which states virgin CRMA performs poor in workability, and as WCO is added, which will be improved in fatigue resistance and remains in an appropriate high-temperature PG grade.

Table 1. Performance indexes of CRMA and CRMA4WCO

Items	CRMA	CRMA4WCO
Viscosity at 135°C, Pa·s	10.50	5.06
Viscosity at 165°C, Pa·s	2.91	1.86
PG grading at high temperatures	PG88	PG82
Fatigue temperature at G*·sinδ=5000kPa, °C	16.7	5.8

3.2. Test and calculation methodology

The frequency creep of samples was conducted at a fixed strain of 10% with temperatures of 0°C, 15°C, 25°C, and 35°C using an MCR 702 DSR apparatus. The target angular frequency variables were followed by 30 points from 0.628 rad/s to 110 rad/s. For running the test, all implementation procedures followed by ASTM D7175.

Based on the study objective, complex modulus (G*) over target frequency variables was transformed into corresponding master curves using the Christensen-Anderson model, the components of which, referring to storage modulus (G′) and loss modulus (G″), were followed as the same.

4. Results and discussion

4.1. Master curves of G at 25°C*

Master curves of G* of CRMA and CRMA4WCO at 25°C with increasing frequency are plotted in Figure 1. As depicted, the G* value of CRMA decreases to some degree with the incorporation of 4%WCO, which at 10rad/s stipulated by Superpave specification (ASTM D7175), changes from 2.9×106 Pa to 1.0×106 Pa, remaining the same magnitude. The result demonstrates that the deformation resistance of CRMA will be somewhat reduced as the limited proportion of WCO is added, which, however, for resulting binder, will not too damaged to bear the outside forces. The reason depends upon that maltene components of bitumen can be further supplemented by WCO with viscosity reduction of integrated binder.

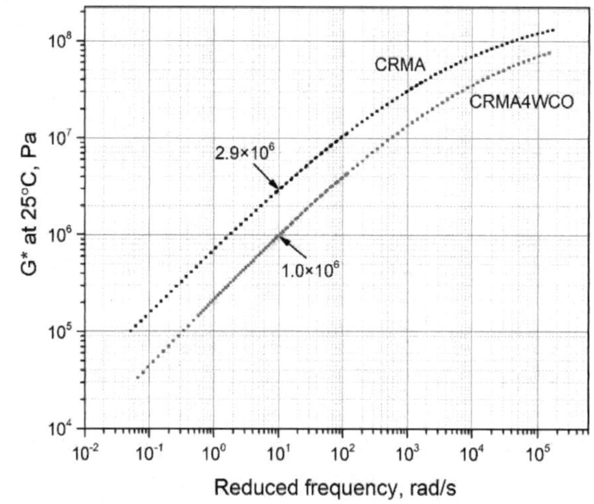

Figure 1. Master curves of G* of binders over frequencies at 25°C

4.2. Determination of viscoelastic transition

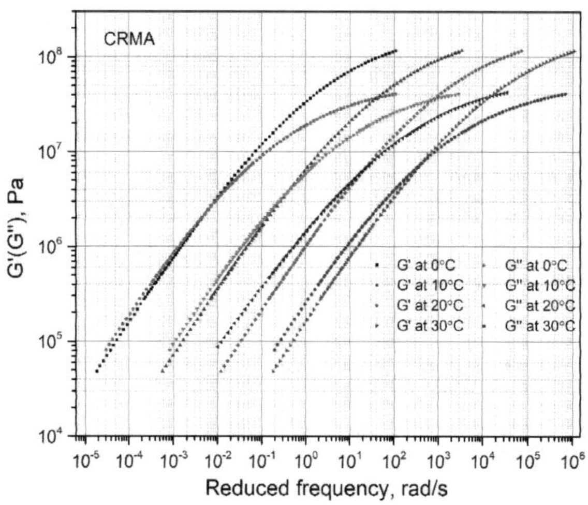

Figure 2. Master curves of G′ and G″ of CRMA binder over frequencies

Figures 2 and 3 show master curves of G′ and G″ of CRMA and CRMA4WCO over frequencies at different temperatures. As plotted, it is well believed that the frequency at intercrossing point of each binder moves towards higher frequency as the temperature increases. The result indicates that increasing temperature needs much higher frequency applied to promote the rheological performance of the binder changing from the viscous behavior to the elastic behavior.

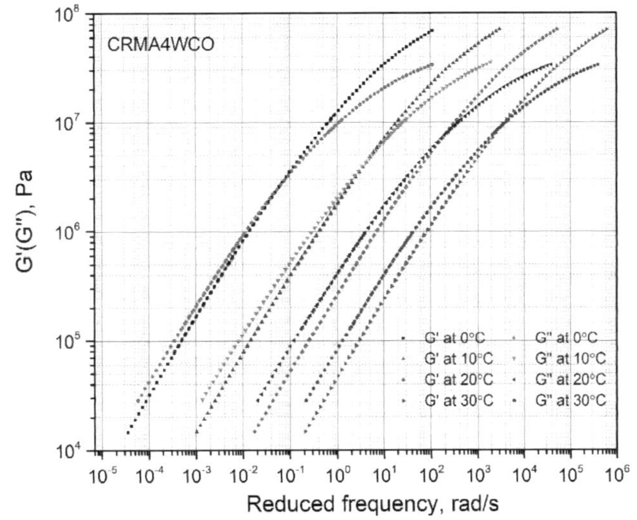

Figure 3. Master curves of G′ and G″ of CRMA4WCO binder over frequencies

4.3. Temperature-frequency dependency of viscoelastic transition

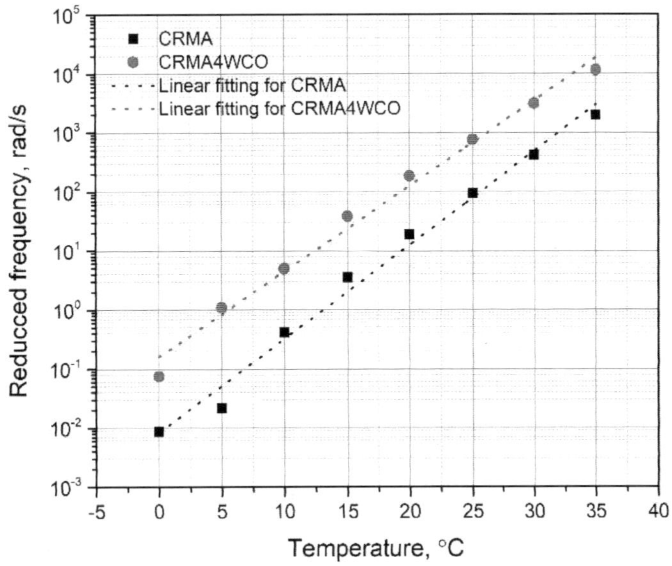

Figure 4. Temperature-frequency dependency of viscoelastic behavior of binder

Figure 4 displays the model values and fitting curves showing the temperature and frequency at viscoelastic transition of binders based on the above-mentioned processing method. Significantly for CRMA4WCO, it needs higher frequency to render the change from viscous behavior to elastic behavior when compared with CRMA. According to the model values at various temperatures, both are fitted in a fitting method as presented in Table 2. It is found that both correlation coefficients, referring to R^2, equal to 0.987, approaching to 1.0, which indicates that the fitting plots are in close relation with the model values and can well reveal the final results in reality. In view of this model formula, the temperature and frequency at viscoelastic transition of binders can be easily determined. Considering the reference frequency (10rad/s) required by Superpave specification, it can be accordingly calculated with the temperatures of CRMA and CRMA4WCO followed as 19.4°C and 12.4°C, respectively. The result suggests that the incorporation of WCO into CRMA will cause a decreasing temperature of viscoelastic transition, and as the temperature raises surpassing the changing temperature of viscoelastic behavior, the major behavior of the binder intends to the viscous behavior.

Table 2. Linear fitting models for determining the viscoelastic transition of binders

Item	Linear fitting model	Correlation coefficient (R^2)	Remark
CRMA	$\log(\omega) = -2.082 + 0.159T$	0.987	ω refers to the respective reduced frequency at viscoelastic transition, rad/s;
CRMA4WCO	$\log(\omega) = -0.798 + 0.145T$	0.987	T refers to the temperature at viscoelastic transition, °C.

4.4. Reliability analysis

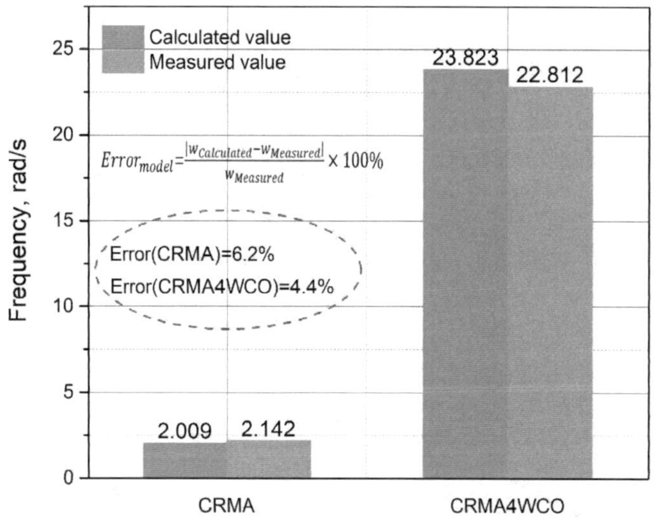

Figure 5. Reliability analysis between calculated and measured values

In consideration with the possible frequency at viscoelastic transition being normally applied around 10Hz (approximately 63rad/s), the frequencies between calculated and measured values at target temperature of 15°C are comparatively analyzed to verify the reliability of the model based on the errors calculated (Figure 5). Obviously, both differences between calculated and measured values for CRMA and CRMA4WCO are less than 1.1 rad/s, while their respective errors are located at 6.2% and 4.4%, which are deemed as acceptable bias for reliable analysis of using the Christensen-Anderson model.

5. Conclusions

This study employed WCO to improve the workability of high-viscosity CRMA, in which the viscoelastic behaviors were analyzed by outputting master curves through Christensen-Anderson model. The findings are drawn as follows. Incorporating 4%WCO into CRMA causes the decrease of complex modulus, but the resulting binder still remains effective in resistance to deformation for general application. Viscoelastic transition of binders needs much higher frequency to be applied when the environmental temperature raises, which to some extent, equals to the effect of incorporating WCO. Once the temperature arrives at the viscoelastic point of binders, they mostly intend to the viscous behavior, and if WCO is incorporated, this behavior will be advanced to a lower temperature. Finally, small errors between values of measured and calculated results are limited within acceptable bias for application reliability of Christensen-Anderson model.

Acknowledgments
This work is financially supported by the Natural Science Foundation of China (Nos. 51808517) and the Science Foundation of Wuhan Institute of Technology (K201932).

References
[1] Leng, Z., Padhan, R. K., & Sreeram, A. (2018). Production of a sustainable paving material through chemical recycling of waste PET into crumb rubber modified asphalt. *Journal of Cleaner Production*, **180**, 682-688.
[2] El-Naga, I. A. & Ragab, M. (2019). Benefits of utilization the recycle polyethylene terephthalate waste plastic materials as a modifier to asphalt mixtures. *Construction and Building Materials*, **219**, 81-90.
[3] Yu, H., Chen, Y., Wu, Q., Zhang, L., Zhang, Z., Zhang, J., Miljković M. & Oeser, M. (2020). Decision support for selecting optimal method of recycling waste tire rubber into wax-based warm mix asphalt based on fuzzy comprehensive evaluation. *Journal of Cleaner Production*, **265**, 121781.
[4] Wang, H., Liu, X., Apostolidis, P., Erkens, S., & Skarpas, A. (2020). Experimental investigation of rubber swelling in bitumen. *Transportation Research Record*, **2674**(2), 203-212.
[5] Venudharan, V., K. P. Biligiri, J. B. Sousa & G. B. Way. (2017). Asphalt-rubber gap-graded mixture design practices: a state-of-the-art research review and future perspective. *Road Materials and Pavement Design*, **18**(3), 730-752.
[6] Mohajerani, A., Burnett, L., Smith, J. V., Markovski, S., Rodwell, G., Rahman, M. T., Kurmus H., Mirzababaei M., Arulrajahc A., Horpibulsuk S. & Maghool, F. (2020). Recycling waste rubber tyres in construction materials and associated environmental considerations: A review. *Resources, Conservation and Recycling*, **155**, 104679.
[7] Ma, J., Hu, M., Sun, D., Lu, T., Sun, G., Ling, S., & Xu, L. (2021). Understanding the role of waste cooking oil residue during the preparation of rubber asphalt. *Resources, Conservation and Recycling*, **167**, 105235.
[8] Yi, X., Dong, R., & Tang, N. (2020). Development of a novel binder rejuvenator composed by waste cooking oil and crumb tire rubber. Construction and Building Materials, 236, 117621.
[9] Xu, X., Leng, Z., Lan, J., Wang, W., Yu, J., Bai, Y., Sreeram A. & Hu, J. (2020). Sustainable Practice in Pavement Engineering through Value-Added Collective Recycling of Waste Plastic and Waste Tyre Rubber. *Engineering*. (In press)
[10] Christensen, D. W., and D. A. Anderson. (1992). Interpretation of dynamic mechanical test data for paving grade asphalt cements. *Journal of the Association of Asphalt Paving Technologists*, **61**, 67-116.
[11] Anderson, D. A., D. W. Christensen, and H. Bahia (1991). Physical properties of asphalt cement and the development of performance-related specifications. *Journal of the Association of Asphalt Paving Technologists*, **60**, 437-475.

AUTHOR INDEX

Institute of Physics
Dirac House, Temple Back
Bristol BS1 6BE UK

ISSN: 1757-8981
ISBN 978-1-7138-3431-1

CURRAN ASSOCIATES INC.
proceedings.com